胡同门楼建筑艺术

The Architectural Art of Hutong Gate Building

（增订版）

（Revised Edition）

李明德　著

李海川　摄影

Written by Li Mingde

Photographed by Li Haichuan

中国建筑工业出版社

CHINA ARCHITECTURE & BUILDING PRESS

胡同巷景尽收眼底
保存旧貌功不可没

舒乙

本书以胡同中四合院门楼，及其门头、门枕石、上下马石、影壁等造型艺术为例，图文并茂地展示其高超的建筑艺术和丰富的民族文化。在门楼建筑上，装饰着砖雕、石雕、木雕，造型多样，画面丰富，工艺高超。虽历经沧桑，至今仍保留下众多的艺术精品，为研究、考证胡同文化提供了生动的依据。

增订版前言

北京是一座世界闻名的文化古都，有着三千多年的建城史。

在丰富悠久的历史文化中，胡同建筑占有重要地位。据史料记载，一些古老街巷形成于元至元十三年（1276年），经历明、清，沿袭至今，大部分建筑及胡同的位置均少有大的变化。

本书以胡同中四合院门楼建筑，及其门头、门枕石、上下马石、影壁等造型艺术为例，通过简要的文字介绍，结合多幅照片，展示其精美的建筑艺术和丰富的民族文化。在门楼建筑上，装饰着砖雕、石雕、木雕，其造型多样、画面丰富、工艺高超。虽历经沧桑，至今仍保留下众多的艺术精品，为研究、考证北京胡同文化提供了生动的依据。每当人们穿街过巷，留心观赏那一座座造型优美的门楼和装饰在上面的砖、石雕刻精品，会得到高度的艺术享受。

北京在前进。在城市现代化正加速进展的时刻，人们对胡同的历史、文化、民俗却有着深深的眷恋之情。本书展示着古老胡同中门楼建筑的画面，从历史和文化艺术角度进行分析介绍，并对所附百余幅照片加以文字说明，以供广大读者研究欣赏，从而进一步认识其历史价值，弘扬它辉煌的一页。

在此增订版中，增加了对南锣鼓巷地区胡同中的老宅、名园、古迹等较为全面的介绍，并附有《南锣鼓巷地区揽胜全图》。结合书中所述的门楼建筑，在南锣鼓巷地区都可以找到可考证的实物。另附有《砖雕艺人邓久安先生访问记》，可让读者对砖雕艺术、砖雕匠人有更加全面的了解。中央文史研究馆馆员舒乙先生在百忙中抽时间为本书题写书名及题词，在此深表谢意！

the Preface of Revised Edition

With a history of more than 3000 years, Beijing is a world-renowned cultural city.

In its rich and prolonged history and culture, construction of alleys or Hutongs occupies an important position. According to historical records, some of its ancient streets and lanes took shape during the *Yuan* Dynasty (beginning from 1276 A. D.) and were carried on to the present through the Ming and Qing dynasties. Most of the buildings and locations of the alleys have not undergone much changes.

Taking the gates of some quadrangles in Beijing's alleys and the decorative arts on the gates, gate lintels, gate piers stepping stones for mounting or dismounting from a horse and screen walls as examples, this book attempts to show the superb, architectural art and folk culture of the gates through brief words and many colored pictures. The varied, rich and fine brick, stone and wooden carvings on the buildings of the gates remain a treasure of works of art today despite many vicissitudes. They provide vivid evidences for the study and textual research of the culture of Beijing alleys. Whenever people walk through the alleys and look at the beautifully-constructed gates and the decorative brick and stone carvings on them they feel they are enjoying a high degree of artistic beauty.

Beijing is fast advancing toward modernization. Still people cherish a profound love for its alleys with their history, culture and folk customs. With a historical and artistic analysis and introduction of the gates buildings in the alleys and with over a hundred illustrated pictures, this book will enable our readers to fully appreciate and further realise their historical values and to carry forward their brilliant aspects.

There are many new contents in this revised edition, including more captions and illustrations about old residences, well-known private gardens, and historic sites in the *Nanluogu* Lane Hutong area. The more valuable content is to add an exclusive interview to an old folk artist named Deng Jiuan of brick carving who passed away many years ago.

目 录

Contents

门 楼

Gate Building

早在元代扩建大都时，北京即遵循《周礼·考工记》中所规定的“左祖右社、面朝后市”的设计原则，对城市的街巷按统一的标准来兴建，城内分五十坊，坊与坊之间形成平直的大街小巷。据明《永乐大典》所辑《析津志》载：大街宽二十四步、小街宽十二步、胡同宽六步（一步为1.54m）。大的街巷为南北方向，一般胡同则沿着南北大街的东西侧顺序排列。明永乐四年（1406年）在元大都基础上又进行了扩展。历经明、清两代乃至民国和新中国成立以后，北京大部分胡同的布局建置均无太大的变化。今天，北京胡同有4000余条。民间流传着顺口溜儿：“有名的胡同三千九，无名的小巷赛牛毛。”一些老街巷，仍保持着原位置和名称。如地安门东大街的南锣鼓巷地区就是很好的实例。元代这里称“昭回坊”与“靖恭坊”。南北的主街、东西对称的十余条胡同形成了蜈蚣状，俗称“蜈蚣巷”。

在北京的胡同里，四合院的门楼是最引人注目的建筑。那一座座造型各异、大小不等的门楼，展示着高超的营造艺术和京城悠久的传统文化。胡同内不同类型的四合院门楼很齐整，一些老宅院门前保存着上、下马石，大门两侧还配有造型精美的抱鼓石，以及影壁、老槐树等景观，其中部分胡同还是沿用原来的名称。这是胡同里古建筑与民俗文化的真实写照。在清代鼓词说唱中，对宅第的门楼如此描述：“青水脊的门楼安着吻兽，两边有两溜拴马的桩。上马石下马石分为左右，铁丝儿的灯笼挂在当央。”北京民间还流传着“宅子老不老，要看槐树大和小”的谚语。其意是通过对四合院门楼前槐树的树龄来验证古老门楼建筑的年代，这还是有说服力的。因为一般大四合院门楼前左右各种槐树一株，其种植是与建门楼同时进行的。四合院不仅是中国古建筑的典型，门楼的样式、规格也因宅第主人的身份而区分。采用坐北朝南的位置，在敞亮的院子里，北面为正房，东西为厢房，南屋称为倒座，门楼大部分均建在院东南角。按古代风水学的要求，“乾宅巽门”为最佳位置。“巽”指东南，故门楼建在四合院的东南方，以示吉利。北京古老胡同里现存的四合院门楼多为清代所建，至今有一二百年的历史，也有极少数是明代的建筑。这些年代悠久、造型优美、不同规格的门楼，包括其门头、戗檐、门簪、门墩……的装饰艺术，都有着高度的研究、欣赏价值，为今天我们了解北京胡同建筑历史及文化、民俗提供了活化石，非常难能可贵。

Back to the Yuan dynasty when Dadu (present-day Beijing) was being expanded, the designing principle followed was "to build the ancestral temple on the left side and site of sacrifices to the God of land on the right side, with the court in front and market at the rear" laid down in *Zhou Li* or the *Book of Rites of the Zhou Dynasty*. The city's streets and lanes were constructed according to a unified standard. There were 50 Fang or portions of land, with straight streets and lanes between them. According to the Yongle Canon compiled during the Ming dynasty, an avenue was 24 Bu(one Bu is equivalent to 1.54 metres) wide, a street was 12 Bu wide and an alley 6 Bu wide. Big streets and lanes were south-north oriented, ordinary alleys ran from east to west on the flanks of the streets. Further expansion was carried out on Dadu in the fourth year of Yongle reign of the Ming dynasty (1406 A. D.). From that time on down through the Ming, Qing dynasties, the Republic of China (1912~1949) and post liberation days the layout of the great part of Beijing has not been changed much. Today Beijing has over 4000 alleys. A popular doggerel goes, "Well-known alleys number 3900 and unknown alleys are countless just like cow hair ." Some old streets and lanes retain their orginal locations and names. For instance, the Nanluogu Xiang district of the Di'anmen Dongdajie (street). During the Yuan dynasty it was called Zhaohui Fang and Jinggong Fang. The north-south oriented avenue and over a dozen symmetrical west-east oriented alleys present the shape of a centipede, hence popularly known as Wugong Xiang or the centipede lane.

The most attractive buildings in Beijing's alleys are the gates of quadrangles. Of various shapes and sizes these gates display China's superb architectural art and Beijing's long-standing traditional culture. The lane boasts rows of neat quadrangles with their variegated shapes of gates. Some time-honored dwellings still preserve stepping stones for mounting or dismounting from a horse in front of their gates, exquisitely skulpted Drum Stones on both sides of their gates as well as screen walls, locust trees and other scenic sights. Part of the alleys still use their original names. These give a true portrayal of ancient buildings and folk customs of Beijing's alleys. In story-telling and ballad-singing of the there was such description of the gates of Beijing quadrangles: "Skulpted animals sit on the bluish ridges over the gates, on two sides are posts for tethering horses, on the right amd left are stepping stones for mounting or dismounting from a horse and lanterns are hung in the centre." A folk saying runs, "To determine the age of a residence, just look at the size of the locust tree." It means that you can determine the age of the residence through the age of the locust trees planted in front of its gate. Generally a locust tree is planted on either side of the gate. They are planted at the same time when the residence is constructed. Quadrangles are a typical example of ancient Chinese buildings. More than that the styles and standards of their gates vary in keeping with the social status of their owners. They were generally built facing south with spacious and well-lighted courtyards. The principal room faced south, wing rooms were on the west and east sides and the room facing north is called Dao Zuo Fang or reversely-set room. Most of the gates were opened on the south-east corners of the courtyards. According to the theory of geomancy, it is the best and most propitous location. Existing quadrangles of Beijing's old alleys and their gates were mostly built in the Qing dynasty with a history of one to two hundred years. A small number of them were built in the Ming dynasty. These time-honored, beautifully-shaped and varied styled gates together with their decorations such as lintels, Qiang Yan or eaves, clasps, piers and others are highly valuable for both research and artistic appreciation. They are living fossils providing us with clues to study the history, culture and people's customs of buildings in Beijing alleys.

王府门楼 The Palace Gate

在封建社会中，门楼显示着主人的品位等级。建筑规格及造型装饰是有所区分的。

清代的王府门楼很壮观，一般是坐北朝南。有五间三启门和三间一启门。如：朝内大街的孚王府，系清康熙皇帝第十三子的府第，正门五间，门前左右各立大石狮一个。后海的恭王府正门三间，门前石狮一对。这样保存完整的府第今天已为数不多了。

※

In the feudal society, the gates demonstrated the ranks and social classes of their masters. They were differenciated in constructional standard and sculptural decoration.

The gates of residences of Princes of the Qing dynasty were magnificent. They usually face south with 5-bay across or 3-bay across. For instance, Fu Wang Fu or the residence of Prince Fu, the 13th son of Emperor Kangxi of the Qing dynasty at Chaonei Dajie. It has a 5-bay principal room having a pair of big stone lions on the two flanks of its gate. Gong Wang Fu or the residence of Prince Gong at Hou Hai or Rear Sea has a 3-bay principal room with a pair of stone lions in front of its gate. Such well-preserved residences are not many .

王府门楼，多为五间三启门或三间一启门
The gates of residences of princes are mostly 5-bay across with three pairs of doors or 3-bay across with one pair of doors

广亮大门 The Grand Light Gate

官宦及富户的门楼多为广亮大门，位置在宅院的东南角。其造型高大，门头及戗檐装饰精美的砖雕，台基也高大，门外广亮宽敞。有些府第的广亮大门门楼，左右两侧砌八字粉墙（俗称“撇山影壁”），上为筒瓦，墙壁磨砖对缝。门楼与粉墙之间留有宽阔地面，设置上、下马石一对，其倒座墙上配有四个拴马桩，更显示其门户整体的端庄和严肃。这样的门楼，今天在老胡同中还能见到。但大多数广亮大门前是不砌八字粉墙的，更不能设置上、下马石。

※

The dwellings of official and wealthy families mostly have Guang Liang gates located on the southeast corners of the houses. These gates are big and tall, their lintels and eaves are decorated with exquisite brick carvings. They also have high steps and spacious areas outside. Some of these gates are flanked with white -washed walls shaped like the Chinese character“八”, (penerally known as Pie Shan screen walls). They are topped with tiles and linked up at the back. Broad spaces are left between the gates and white-washed walls where a pair of stepping stones for mounting and dismounting from a horse are installed. There are also 4 posts for tethering horses. All of these add dignity and solemnity to the gates. Such gates can still be found in old-time alleys today. But most of the Guang Liang Gates do not have white-washed walls, let alone stepping stones for getting on or down from a horse.

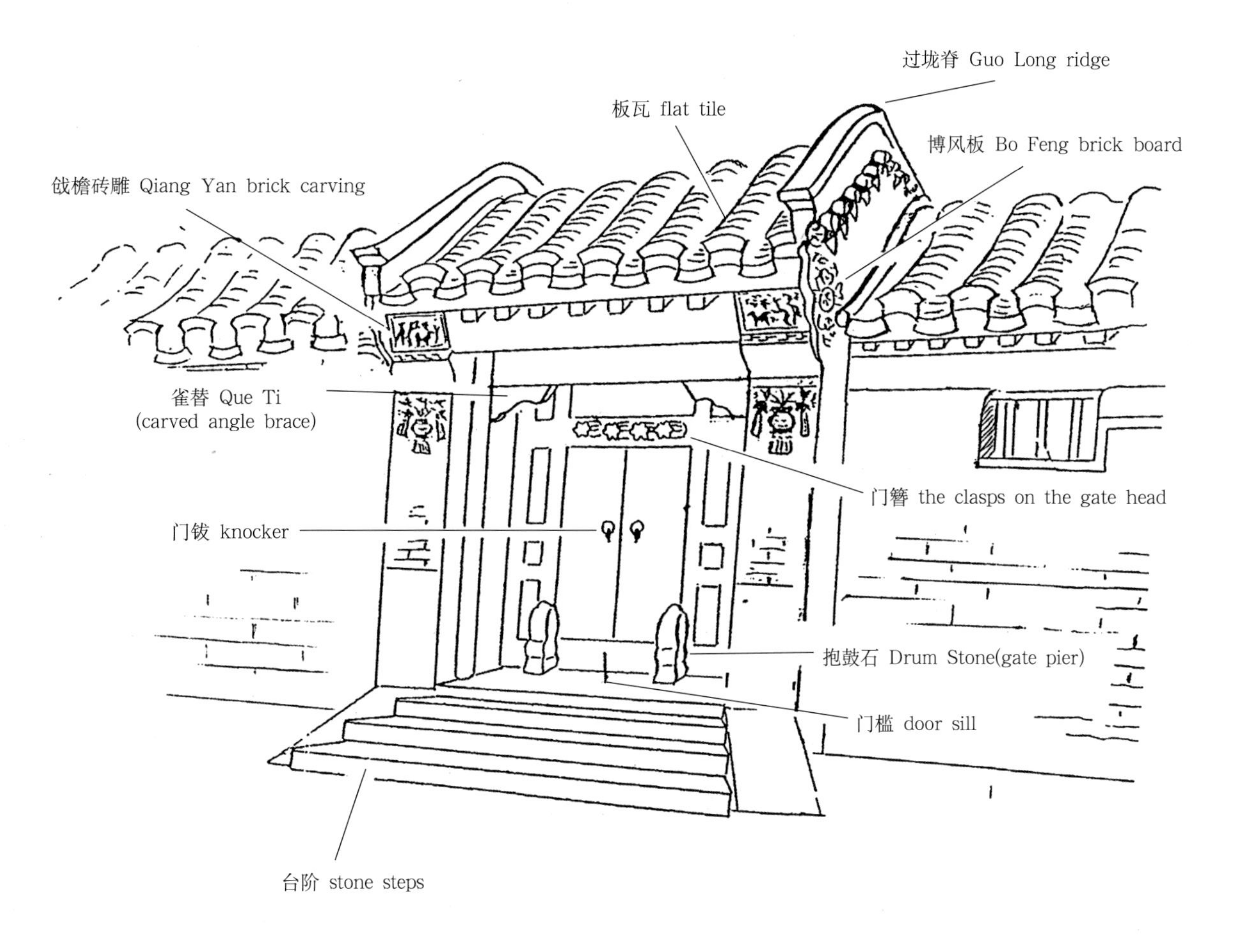

广亮大门门楼建筑结构图

The construction of the Guang Liang Gate

广亮大门门楼造型精美，砖雕工整，显示出宅主人的品级地位。

A well-preserved gate of a large quadrangle. Exquisitely-designed brick carvings show the social position of the house owner.

如意门 Ruyi Gate

一般中小型四合院多为如意门。在北京数千条胡同中，如意门楼的数量很多。其建筑特点是大门设在外檐柱间，门框两侧砌砖墙。门楣上有较精细的砖雕图案，门楣与两侧砖墙交角处砌如意状的砖饰，表示“吉祥如意”，故称如意门。街巷中的如意门楼多为清代建筑，也有部分是民国期间所建的。大型如意门，门楼占一整间；中型如意门，门楼占半间；小型如意门，门楼占半间。在为数众多的如意门中，其建筑造型是多彩多姿的。有的砖料上乘，砌工考究，门头栏板及戗檐的砖雕图案内容丰富，门簪上的木雕工艺精巧，有文字或花卉图案，有的门上还钉有对称的铁皮图案。这些建筑上的装饰都反映出我国古代劳动人民的智慧，其中精品有着高度研究价值。

※

Ordinarily medium-sized and small quadrangles have Ru Yi Gates. Their number is the greatest in several thousands of Beijing alleys. They are opened between outer columns and have brick wall on both sides. On their lintels, there are fine patterns of brick carvings. The juncture of lintels and brick walls deco-rative Ru Yi-shaped bricks implying auspiciousness. Thus people call them Ru Yi Gates. Most of the Ru Yi Gates of Beijing's streets and lanes were built in the Qing dynasty, part of them were constructed during the Republic of China (1912~1949). Big Ru Yi Gates are one-bay across, medimm-sized ones occupy more than half bays and small Ru Yi Gates take up half a bay. Buildings of the numerous Ru Yi Gates are rich and colorful in architectural design and style. Some have meticulously-laid, fine-quality bricks, their panels and eaves bear a great variety of carved brick patterns and their gate clasps are decorated with fine wood carvings with words or floral patterns. Some Ru Yi Gates are set with symmetrical patterns of iron sheets. All these constructional decorations show the wisdom of the working people in ancient China. Exquisite works among them are of great value for research work.

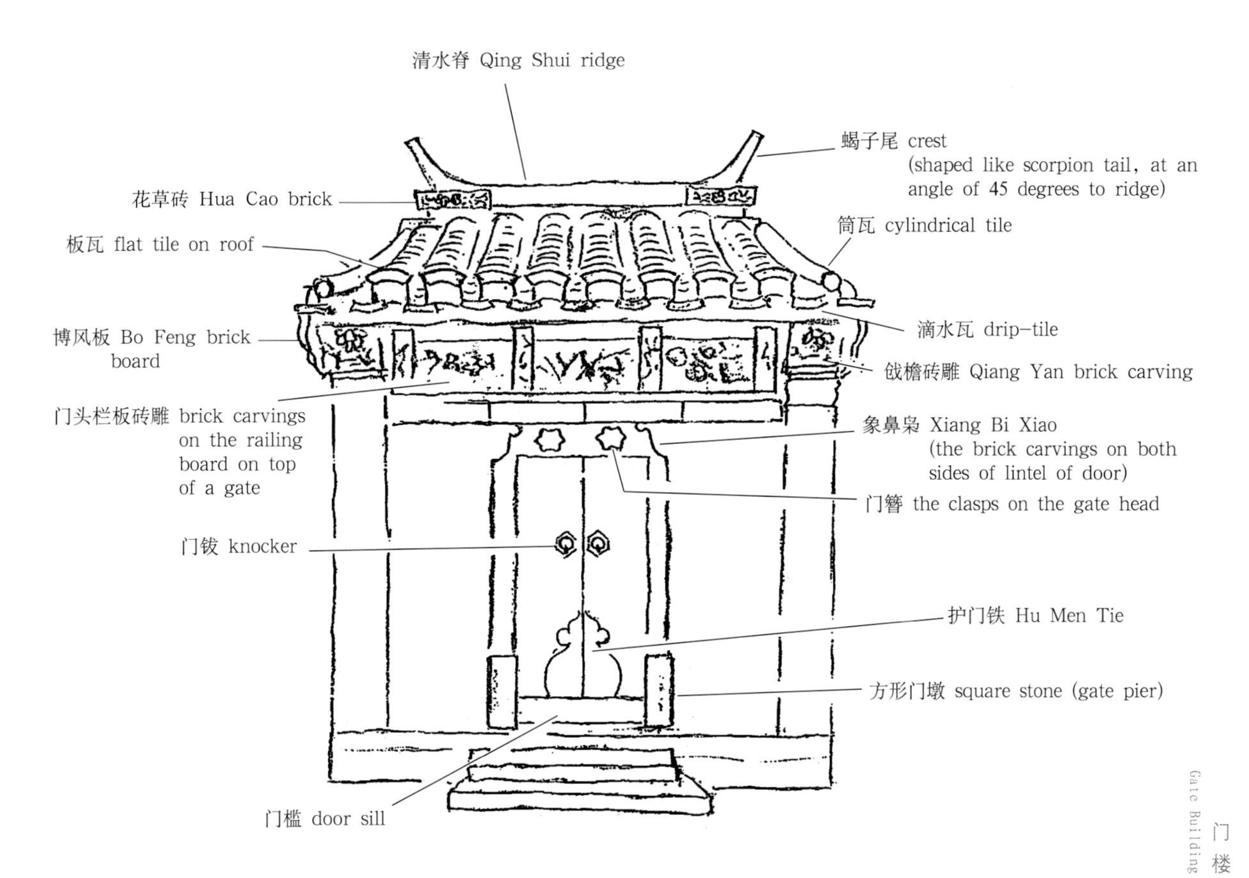

如意门门楼建筑结构图
The construction of the Ru Yi Gate

如意门门楼
Numerous Ru Yi Gate

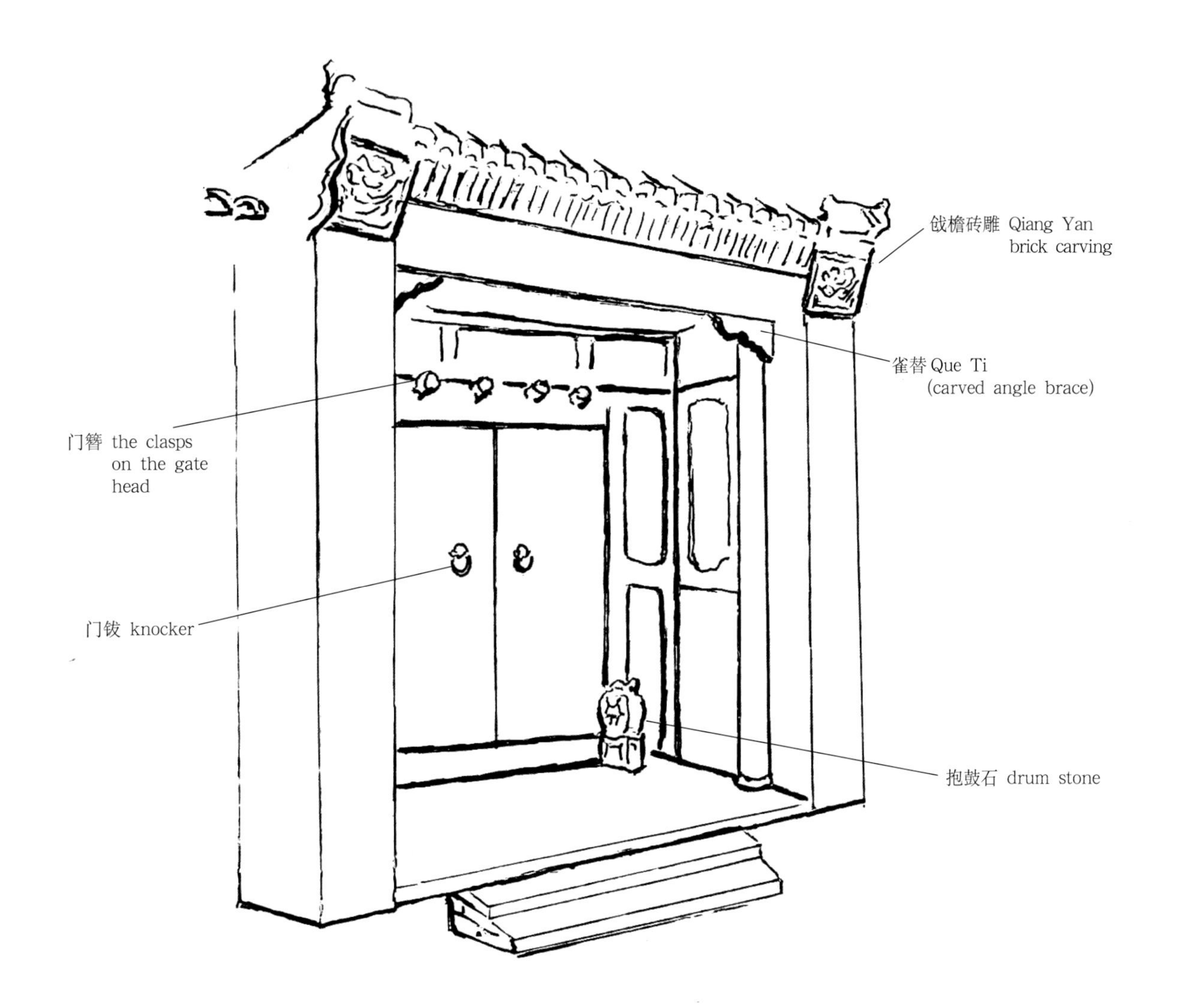

金柱大门门楼建筑结构图

The construction of the Jin Zhu Gate

金柱大门
Jin Zhu Gate

9

蛮子门 Manzi Gate

蛮子门等级比金柱大门略低。此种大门的门口又向前推一步，立在了前檐柱的位置。它的构造如门板、槛框、门枕石都与金柱大门差不多，但没有雀替。一般住在此院的主人没有官品，大多为南方经商的人。此种门名称不雅，顾名思义，有南方的建筑式样。

※

Manzi gate, another type of gate building of residence in Hutong (old alleys), has not gateway with its panel set between outer columns. Its building construction like Jinzhu gate buildings. As much as we know, the owners of the residence of Manzi gate alway be businessmen from South China. So this style of gate like the gate buildings in South China.

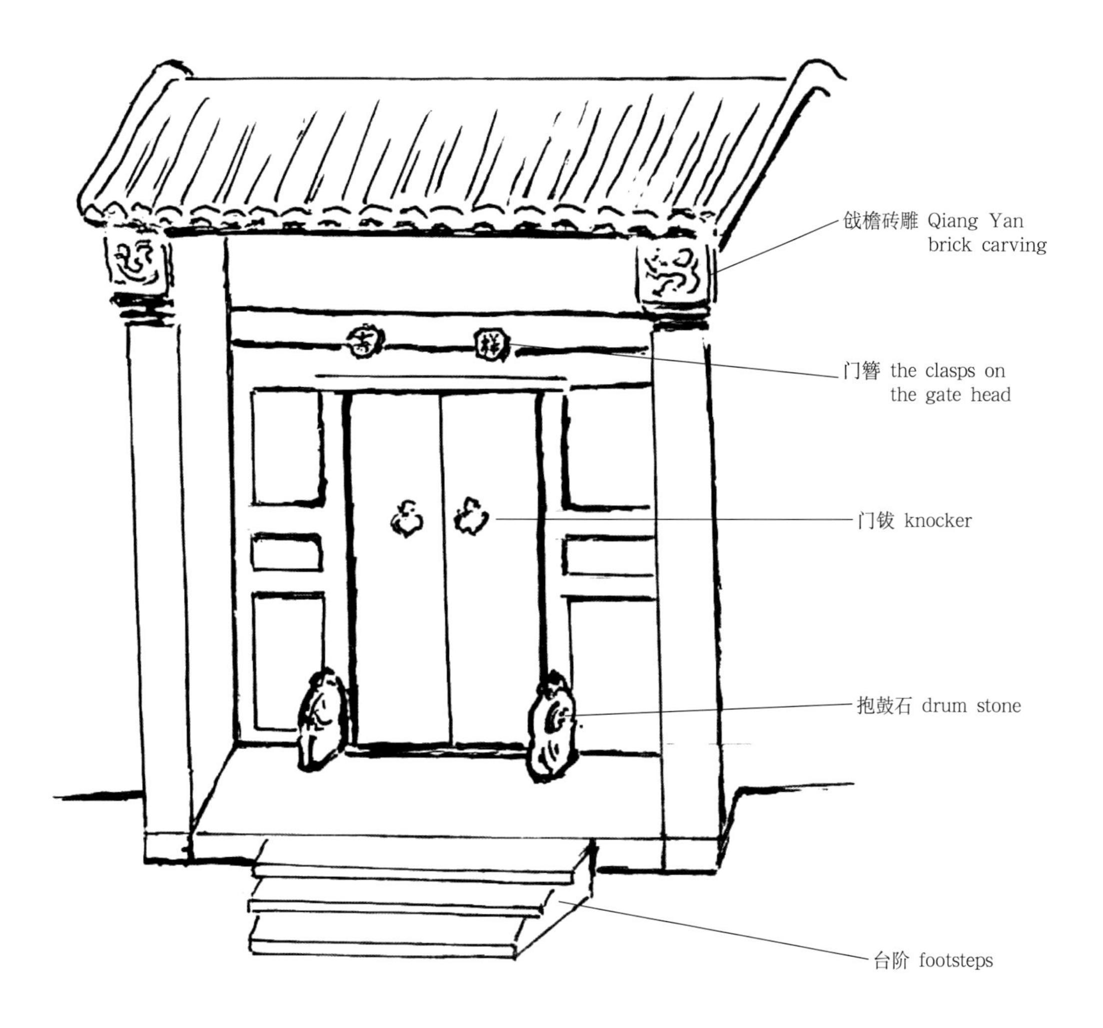

蛮子门门楼建筑结构图
The construction of the Man Zi Gate

蛮子门，大门的门扉安装在外檐柱间就没有大门洞了，是广亮大门的另一种形式

Man Zi Gate is another type of Guangliang Gate. With its panels setting between outer columns , it has no gateway.

蛮子门
Man Zi Gate

随墙门 Sui Qiang Gate

胡同里还有为数不少的墙垣式门，其规格较小，门楼与院墙相连，故又称“随墙门”一般较小的四合院或三合院多采取这样的建筑形式。在胡同里，老北京平民百姓所居住为数最多的是随墙门，其造型很简易，然而又最为普遍。有的门墙用碎砖头和泥灰砌成，北京人称这类碎砖墙为“核桃酥”，比喻其建筑质量之差。

※

There are also a good number of wall-typed gates, comparatively smaller. Linked with walls of courtyards these gates are generally adopted by smaller quadrangles or San He Yuan, threeroom compounds. Besides there are Sui Qiang Gates adopted by the majority of old Beijing residents. They are simple and widespread. Some Sui Qiang Gates are built of broken pieces of bricks and marl. Beijingers call these gates “Walnut shortbreads”, a metaphor insinuating their poor quality.

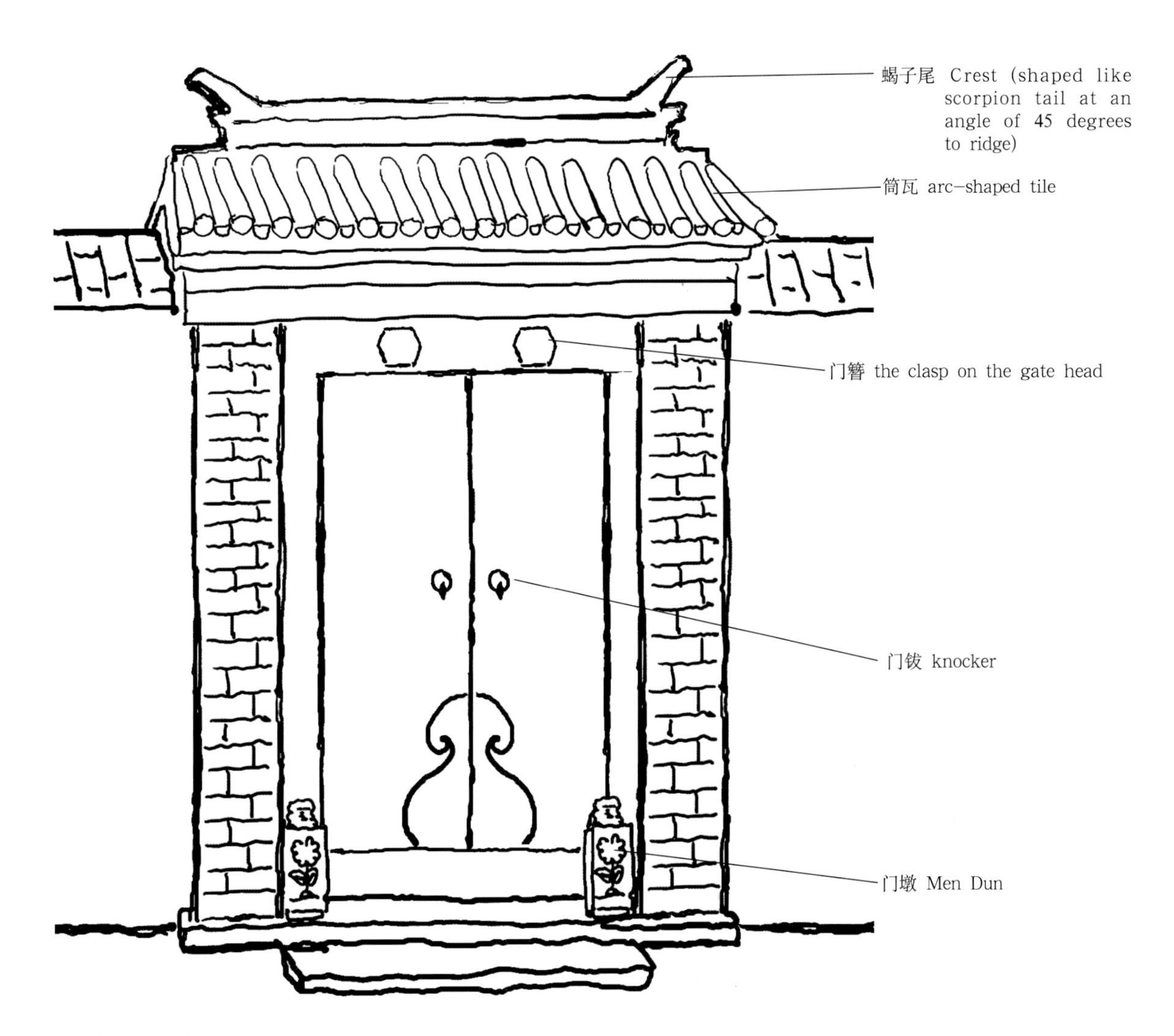

随墙门门楼建筑结构图

The construction of the Sui Qiang Gate

随墙门
Sui Qiang Gate

西洋式门楼

The Gate of Western Style

清末至民国时期，北京胡同中出现了不同类型的西洋式门楼，这是中国传统北方民居建筑受西方文化影响的结果。有的门楼呈现了中西合璧的新造型。据说最早的西洋式门楼出现在圆明园，故北京人也把这样的门楼称为“圆明园式门楼”。

西洋式门楼主要在门头和大门两侧墙壁上有西式建筑风格的装饰，但大部分又保留了中式建筑原有部分的风貌。例如有的大门左右保留着方式门墩儿，门墩儿上雕刻有“佛八宝”图案，还有的门楼旁保留着有石雕的拴马桩。

这类既有传统中式特色又有西洋式建筑风格的门楼得到了人们的认可，有很大的欣赏价值。

Different gates buildings of western style appeared during the late Qing Dynasty in Beijing, for the Chinese traditional buildings were affected by the western culture. There were some gate buildings as the combination of Chinese and Western features in that time in Hutongs. It is said that the like the gates of the style named ‘Yuan Ming Yuan’ .

The feature of this style of gate buildings is the western style of decoration, but they inherited from the gate buildings decoration of the Chinese traditional style. Today, there are still some different traditional contents of patterns on the squire-shaped gate piers like the patterns of 'Buddha treasures'. There are a few old stone-made posts to tie horses which are preserved near the gate buildings.

In a word, these gate of western style draw more and more attention to them. Their value is beyond us.

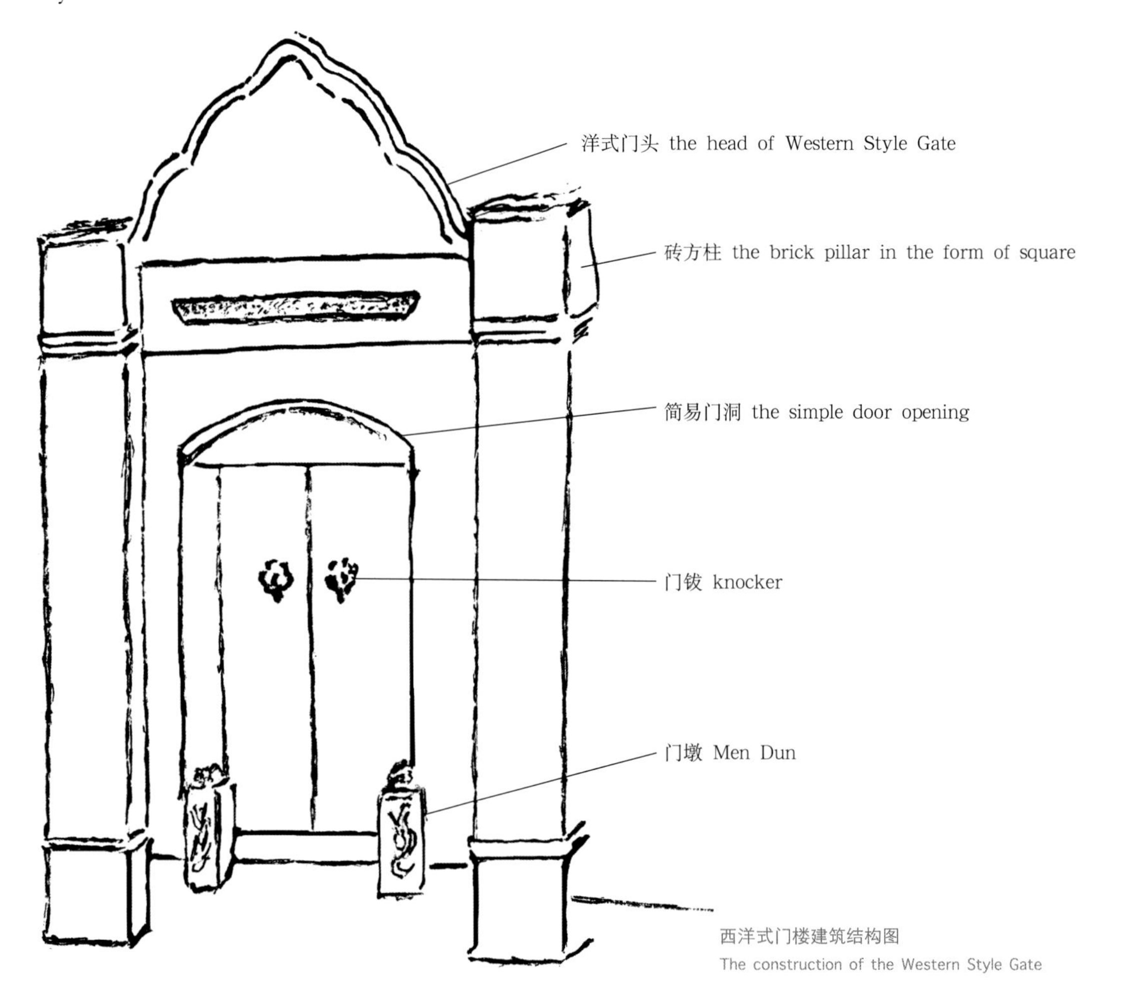

西洋式门楼建筑结构图
The construction of the Western Style Gate

西洋式门楼

the Western Style Gate

西洋式门楼
the Western Style Gate

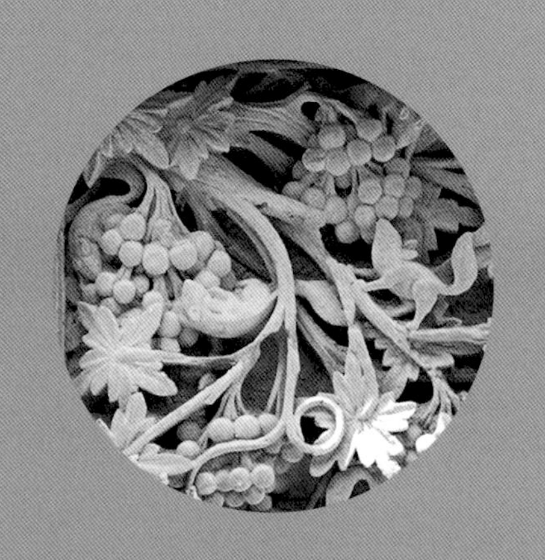

门楼砖雕

Brick Carvings on the Gate Building

北京胡同四合院门楼上的砖雕、石雕装饰，有着高度的观赏价值，独具艺术光彩。门头上的栏板及两旁戗檐上的砖雕图案内容丰富，雕刻技法高超。虽经过百年以上的风雨侵蚀，这些精美佳品至今仍保存完好，与整体门楼建筑结合得恰到好处，形成极好的装饰效果。

追溯北京砖雕艺术的历史，以明清时代最为兴盛，那时府第、会馆、民居的四合院门楼上装饰砖雕、石雕很普遍，用于门头、门墩儿、屋脊等处。现存的一些老街巷门楼上的砖雕，多为清代工匠所雕，也有部分是民国期间的砖、石雕品。工匠们根据门楼宅第的不同，采用薄肉雕、浮雕、透雕和线刻等多项高难的技法，在门楼、门头、戗檐、门墩儿上进行图案设计，按照图样的尺寸去烧制澄浆泥砖，在砖上雕刻。其中以浮雕最具特色，有些画面的精彩部分，则单独制作，后再镶在砖面上，如花朵、兽头……还有的部分运用透雕技术，表现出的图案效果更佳。之后再用拼接的方式，如两拼、四拼、六拼、八拼等完成门头的整个砖雕装饰。画面上呈现的人物、花卉图案的完成，要靠工匠的刀工技巧，一般雕完后不再另行打磨，几块砖雕拼砌在一起，就形成了巨幅的佳品了。今天，北京会这门砖雕技术的老艺人已寥若晨星。

Brick and stone carvings on the gates of quadrangles in Beijing alleys are highly valuable for ornamental purpose and possess unique, artistic splendor. The carved brick patterns on the lintels and on Qiang Yan or eaves on both sides are not only rich in content but also superb in carving skill. Though weatherbeaten for over a hundred years the cream of them are still preserved. Well matched with the whole gate buildings they achieve a good decorative effect and are loved by the people. The history of the art of Beijing's brick carving can be traced back to the Ming and Qing dynasties when it was flourishing. At that time it was popular to mount brick and stone carvings on the lintels, piers and roof ridges of the gates of quadrangles of civilian dwellings, guild houses and officials' residences. Most of the extant brick carvings were made by artisans of the Qing dynasty, part of the brick and stone carvings were done during the Republic of China (1912～1949). Applying various highly difficult carving skills such as thin carving, bass relief carving, open work carving and linear carving, the artisans first drew designs on the arch gateways, lintels, Qiang Yan or eaves and gate piers and baked bricks according to the designed sizes and then made carvings on them. The most outstanding is bass relief carving. Attractive parts like flowers, animal heads and others were made separately before they were inlaid on the bricks. Open work carving technique could achieve the best artistic effect. The whole carved brick decoration of the gate heads was accomplished by joining together 2,4,6,8 and more pieces. It depended on the craftsmanship of artisans to complete the patterns of figurines, flowers and plants. Once carved, usually no further polishing was needed. The joining together of several carved bricks formed a great master-piece. Today only a small number of masters skilled in such brick carving can be found in Beijing.

门楹砖雕 Brick Carvings on the Gate Lintels

门楹（或称门额）、戗檐上的砖雕精品，有着浓郁的地方特色，是胡同文化史与民俗的写照。细心观赏，能分辨出画面的丰富内涵。砖雕的内容大致可以分为如下几类。

1.吉祥富贵的图案。展示人民对美好幸福生活的向往。以蝙蝠、鹿、喜鹊、仙鹤、麒麟、盘长、如意、磬……分别组成图案，各有其讲法和寓意。有的图案还与其谐音联系在一起，很有讲究。在栏板浮雕蝙蝠（谐“福”音）、鹿（谐“禄”音）、古代乐器磬（“磬”即是喜庆的谐音）。“喜庆吉祥”的图案中，画面上的蝙蝠口叼着磬，表达了主人求喜庆、盼福到的心愿；“福寿绵长”图案中，蝙蝠口叼盘长（佛八宝之一）其图案中绸带不封口，表示佛法无边，其寓意是福寿绵长；“鹤鹿同春”图案，表示延年益寿。

2.展示花卉的图案。如梅花、兰草、翠竹、菊花，这是中国的传统名花，称为“园林四君子”，自古以来就得到文人墨客的吟咏。宋代文学家王安石就留有赞梅花的诗：“墙角数枝梅，凌寒独自开。遥知不是雪，为有暗香来。”松、竹、梅组成图案砖雕，称为“岁寒三友”。还有雕刻石榴、葫芦、葡萄图案的，其寓意是多子多孙、人丁兴旺。

3.“博古”图案。画面内容丰富。将古代器物香炉、玉佩、笔筒、砚台、花瓶……巧妙安排在一组“多宝阁”画面中，表示我国文人的生活。文房四宝与花瓶在一起，寓意四季平安。

4.民间传说与神话故事。经过考查，现在这类人物故事题材的砖雕作品为数极少了。我国古老的“八仙过海”故事，将韩湘子、张果老、李铁拐、曹国舅、蓝采和、汉钟离、吕洞宾、何仙姑的传说，被活灵活现地展现在画面上，是北京砖雕艺术中的佳品。表现形式上，又分“明八仙”和“暗八仙”两种，韩湘子吹笛、曹国舅持玉板高歌、李铁拐祝寿、何仙姑采莲、张果老骑驴……这类人物图案为明八仙。在图案上仅雕刻八仙常佩带的器物，如：韩湘子吹的笛子、曹国舅手中的玉板、蓝采和的花篮、汉钟离的扇子、何仙姑手持的荷花、铁拐李的葫芦、吕洞宾的宝剑、张果老的渔鼓配之以祥云、绸带、荷叶组成图案的，均为暗八仙。京城有一处门头砖雕，名为“竹林七贤”。上有象征吉祥的佛八宝图案，即宝伞、胜利幢、宝瓶、双鱼、莲花、海螺、吉祥结、金轮，更是珍品，展现魏晋年间，阮籍、山涛、刘伶、向秀、王戎等七位文人名士经常游于竹林饮酒赋诗，抚琴吟唱的情景，画面生动感人，是目前胡同砖雕仅存的精品。其他如“麒麟送子”“马上封侯”等图案，内容可谓丰富多彩。

5.亭台楼阁图案。画面多以表现四合院花园建筑景色为主。以亭、廊、假山等景观，突出浮雕特点，展示庭院的情趣。

6.“福”字、“寿”字、云纹等图案。门楼左右上方的戗檐，砖雕图案多为花卉、动物或人物故事。如：雕刻兽中之王狮子的图案，象征着宅主人的武官身份。雄狮戏球的画面更为生动活泼，这里还有太狮、少狮之分。还有浮雕富贵牡丹或牡丹花篮、松鹤同春等花卉图案的。戗檐上的人物砖雕更为罕见。这些图案构图精美、生动，刀工玲珑剔透。

※

With a strong local flavor exquisite brick carvings on the gate lintels (or gate heads) and Qiang Yan or eaves are a portrayal of the history of Hutong (or alley) culture and folk customs. At a close look one can understand the rich contents of their designs. They can be divided into the following categories.

1. Designs of auspiciousness, wealth and nobility express people's yearning for a happy life. Patterns composed of bats, deer, magpies, white cranes, Chinese unicorns, Pan Chang Ru Yi (an

S-shaped ornamental object symbolizing good luck), chime stones, etc. Each has its own meaning and implications. Some patterns have something to do with homophony. In bass relief carving on railing boards, bat (Bian Fu) is homonymous with "福" (Fu) or good fortune, deer pronounced Lu in Chinese with "禄" (Lu) or official rank and chime stones pronounced Qing in chinese with "庆" Qing or jubilation. In the pattern of "jubilation and auspiciousness" the bat (s) holds a chime stone in its mouth, expressing the house owner wish for attaining happiness and good fortune. In the design "infinite happiness and a long life" the bat holds in its mouth a Pan Chang, one of the 8 Buddhist treasures ("Chang" in Chinese means long). For the silk ribbon in the picture not being closed implys that the powers of Buddha are boundless and that the owner can enjoy infinite happiness and longevity. The pattern of "crane and deer" expresses the good wish for a prolonged life.

2. Designs of flowers such as plum blossoms, orchids, emerald bamboos and chrysanthemums. These are all China's famous traditional flowers known as "the four gentlemen of the garden" They have been praised in poems by letters men since ancient times. Writer Wang Anshi of the Song dynasty composed a poem eulogizing plum blossoms. The poem goes like this:Several plum blossoms at a wall corner, blooming lonely in defiance of cold; Knowing from afar they are not snow for their fragrance is wafted in the air. The pattern of pine, bamboo and plum blossom on carved bricks is called "the three companions of cold winter" . Besides there are carved pomegranates, gourdsand grapes implying bearing many children to make the family tree thrive.

3. Designs of ancient relics. Ancient articles such as incense burner, jade plate, brush pot, inkslab, vase……are ingeniously put together to form a picture of "curio shelf" , depicting the life of Chinese scholars. To place the four treasures of the study namely writing brush, ink stick, ink slab and paper with the vase expresses the meaning that things would run smoothly in all four seasons.

4. Folk tales and mythological stories. After investigation only a small number of carved bricks on such subject matters exist today. The best of Beijing's carved brick art are the fairy tale about eight immortals crossing the sea. The 8 immortals are Han Xiangzi, Zhang Guolao, Li Tieguai, Cao

Guojiu, Lan Caihe, Han Zhongli, L Dongbin and He Xiangu. There are two ways of presentation, open and concealed. The open way is to portray Han Xiangzi playing flute, Cao Guojiu singing with jade clappers, Li Tieguai attending a birthday celebration, He Xiangu picking lotuses, Zhang Guolao riding on a donkey so on and so forth. The concealed way is to carve only the articles the 8 immortals usually carry such as the flute of Han Xiangzi, the jade chappers of Cao Guojiu, the floral basket of Lan Caihe, the fan of Han Zhangli, the lotus of He Xiangu, the gourd of Li Tiegua, the sword of L Dongbin and the Yu Gu, a percussion musical instrument made of bamboo, decorated with clouds, silk ribbons and lotuses. Brick carvings on the gate heads entitled, "7 outstanding scholars of bamboo grove" are also gems of art that have been left. It have patterns of the eight Buddhist treasures symbolized auspiciousness: Umbrella, Sheng Li Chong (streamer used in ancient China), Vase, Double Fish, lotus flower, Ji xiang jie (the Chinese Knot), Jin Lun (the pattern shaped like wheel). They describe scenes about 7 men of letters, including Yuan Ji, Shan Tao, Liu Ling, Xiang Xiu and Wang Rong during the period of Wei and Jin who composed and recited poems, played lutes and drank wine while roaming in bamboo groves. The portrayals are vivid and moving. Other designs about folk tales and fairy stories include "The unicorn brings a baby" and "returning the granted title and seal" .

5. Designs of pavilions and chambers. The brick carving's patterns of terraces and towers show the rich garden views of traditional quadrangle yard.

6. Designs of Chinese characters such as for fu meaning good fortune, shou meaning longevity and yun meaning clouds. Carved brick patterns on Qiang Yan over the right and left sides of the gates are largely flowers, animals or characters in novels. For example, the pattern of lion and the king of animals indicates the military rank of the house owner. The picture of lions playing with a ball is lively and vivid. In the pictures there are young lions as well as grown ones. Floral patterns carved in bass relief include peony or peony basket symbolising wealth and rank and "pines and cranes" signifying longevity. Brick carvings of characters on Qiang Yan are a rare sight.

由三羊开泰、博古图、鹤鹿同春图案组成的整体门楹砖雕，局部按3部分拍摄
This is an overall brick carving composed of three parts. The patterns are the curio shelf, three goats, pine and crane.

门楹砖雕“博古图”。下至挂落板配有人物和蝙蝠、花卉图案浮雕。局部按3部分拍摄

The Curio Shelf Picture of brick carving on the upper side of gate (The design portrays Chinese ancient relics, known as “the Curio Shelf”). The design of brick carving under the Curio Shelf Picture is composed of figures, bats, flowers.

现保存完好的雕有博古图的大型如意门砖雕全貌

This is an overall picture of brick carvings on a large Ru Yi Gate found in an old alley, which is decorated with"the Curio Shelf".

老胡同里不同年代、造型多样的门楹砖雕图案

Various patterns of brick carvings on the gate head in old Hutongs.

门头栏板砖雕松、竹、梅、兰图案

A set of brick carvings on the railing board on top of a gate with patterns of pine, bamboo, plum blossoms and orchids.

一组年代久远的人物砖雕

The brick carvings of the figures patterns has a very long history.

如意门门楹为数很少的石板雕，造型古朴、大方，独具特色
The Railing Stone Board on top of Ru Yi Gate is terse, unique and rare.

浮雕多种花卉图案的门楹

The relief brick of the various flowers patterns on the gate head

砖雕精美的宝瓶与花卉的门楹
The brick carving of the vases and flowers patterns on the gate head

透雕“八骏图”及花卉的门楹
The pattern of eight horses and flowers carved vividly

不同图案的门楹砖雕

The brick carvings of different patterns

戗檐砖雕 The Brick Carving on Qiang Yan

清代官宦之家门楼戗檐砖雕，上边浮雕“鹤鹿同春”，下面是“富贵牡丹”花篮图案

These are brick carvings on Qiang Yan (eave) of the gates of official families of the Qing dynasty. The upper part is the bass relief pattern "Crane and deer" while the lower part is the pattern of a floral basket entitled, "Peony, the symbol of wealth and nobility".

戗檐与博风板的砖雕图案

The patterns of brick carvings on the Qiang Yan and Bo feng Ban

戗檐砖雕“狮子绣球”图案，显示宅主人的高贵身份。狮子寓意尊严。在胡同里很少见到，可谓砖雕精品
The pattern, “Lions play with a ball”, indicates the high rank of the house owner. Lion symbolizes dignity. It's rarely seen in the alleys nowadays so it is a gem of brick carvings.

砖雕麒麟图案，是吉祥与智慧的象征

The pattern of "Unicorn" symbolize the wisdom and auspiciousness.

透雕“博古图”，展示宅主人的文化品位
The brick carving of the curio shows the house owners' taste for culture.

广亮大门门楼上一左一右的戗檐以山羊、猴子、花草组成的画面，生动活泼，透雕精美
Brick carving on the Qiang Yan on the both sides of gate head composed of goats, monkeys, glass.

“马上封侯”图案

The pattern named “Ma Shang Feng Hou” (It's the Chinese traditional subject to symbolize granted title composed of horse monkeys)

“犀牛望月”图案

The pattern of a bull viewing the moon

透雕“青竹翠鸟”图案
The open work carving of the bamboos and bird

古朴的“松鹿图”
The pattern of primitive simplicity of pine and deer

如意门门楼戗檐，砖雕篆字“延年益寿”

A relief brick carving on Qiang Yan composed of four Chinese characters “延年益寿” and bats, clouds and others.

葫芦、葡萄、石榴透雕

A open work carving with the patterns of gourds, grapes and megranates

由荷花、菊花、牡丹组成的戗檐砖雕

Brick carving on the Qiang Yan composed of lotus flowers, chrysanthemums. and peonies

由牡丹、翠竹、兰花构成的戗檐砖雕

Brick carving on the Qiang Yan composed of peony, bamboo and orchid.

象鼻枭砖雕 The Brick Carving on Xiang Bi Xiao

清水脊与花草砖砖雕

Qing Shui Ridge and Brick Carving of Flowers and Plants

清水脊与花草砖
Qing Shui Ridge and Huacao Brick

屋瓦 Tile

工精美，受到人们的喜爱。相附在鼓形及方形石墩儿正面及两侧面上的浮雕图案更是花样繁多，匠心独具。门墩儿在大小规格尺寸上是等级森严的。①王府大门的抱鼓石高为0.80m、宽为0.45m、厚为0.30m；②广亮大门的抱鼓石高为0.75m、宽为0.45m、厚为0.30m；③如意门的抱鼓石高为0.60m、宽为0.29m、厚为0.19m；④平民百姓家的门墩儿高0.59m、宽0.25m、厚为0.19m。王府与百姓之家差异很大，正、侧面上的雕刻图案，刻工的精细程度也有所不同。大部分浮雕画面是花卉，如：梅花、菊花、兰草、翠竹，这是我国的传统名花。还有石榴、蝙蝠、玉笛、葡萄、如意等图案。西城一座王府大门的抱鼓石正面及两侧面的浮雕为古代“乐舞人”，图案造型优美，舞蹈姿态生动，给人们以美的享受。在一些老胡同中，还有“福在眼前”（“前”与“钱”为谐音）这类的门墩，正面精雕蝙蝠叼古钱的图案，其寓意“福在眼前，财源滚滚”之意,表示了宅主人的意愿。以上叙述的这几种图案，今天在胡同中细心观赏，还是能见到的。但我们在胡同中考查时，发现大部分石狮子，在“文化大革命”破“四旧”中遭了难，有的四合院门前石墩儿仅存一个完整小狮子，或一对狮子均头尾不全，伤痕累累。今天一些老胡同里还保存着极少部分成对完整无损的小狮子门墩，可谓珍品。

While enjoying the beauty of brick carvings on the gates' heads one should also pay attention to the gates' piers on both sides of the threshholds. In architecture they are called the gates' pillows. They are popularly known as grtes piers. A Beijing doggerel goes: "A little boy sits on the gate pier, he cries for getting a wife; getting a wife for what, to have someone to chat with by light and to have a companion after the light is turned off. The next morning she will braid the piglet for me." This is a good portrayal of children playing in front of the gates of quadrangles and their early relations with gates' piers. As far as stone carving is concerned existing gates' piers of Beijing alleys can be divided into two types. Ancient ones shaped like drums are called Drum Stones while those carved in the recent hundred years are rectagular. The majoriity of carved ptterns on the top of them are small lions. On the front, right and left sides below, there are carved a great variety of patterns in bass relief. Each and every one of them is a shining example of superb stone carving. There are many stories about the small reclining lions on top of the gates' piers. A folk saying goes: the dragon has 9 sons. Though each has a special skill they all haven't become dragons. Ancient China has many tales about the dragon. One of them goes: the dragon has given birth to 9 different species of animals. The eldest is called Pulao who likes making sounds. So he becomes a handle on the bell; the second son is named Qiuniu, he is fond of music and is carved on Huqin, a Chinese musical instrument; the third son's name is Yazi who likes killing and is thus used to decorate swords and knives; the fourth son is named Chaofeng who loves adventures and so he is used to sit high on the roof ridges of pavilions and halls; the fifth son Suanni who likes sitting becomes an animal on Buddha's pedestral; the sixth son, Baxia, popularly being known as tortoise, is capable of bearing heavy things. He serves as the base on which stone tablets or monu-ments stand. The seyenth son, Bi'an is indulged in making lawsuits. Thus he becomes an object of suppression over the gate of prison; the eighth son called Xibi loves literature and is used as undulating clouds to embellish both sides of stone tablets; the ninth son, Jiaotu, is quick to respond and likes closing, so he seryes as a doorkeeper performing the duty of "closing the door after entering" The 9 sons of the dragon each has his own peculiar capability. (According to references from "Legends about Emperor Qianlong") Jiaotu is the ninth son of the dragon. There are records about him in a chapter on the 9 sons of the dragon from a book of the Qing dynasty enti-tled, "Ren Hai Ji" or "On the Sea of Man" Jiaotu is agile and fond of closing, thus serving as a doorkeeper for his master" The carved little stone lions on the gates' pillow stones are one of the dragon's 9 sons. Meticulously carved to present various postures and expres-sions they are beloved by the people. Bass relief patterns on the front, right and left sides of the drum-shaped and square gates' piers are also rich and colorful and uniquely carved. Gates piers vary in size and standard according to the ranks of house owners. 1. Gates piers of Princes' mansions are 0.80metres tall, 0.45 metres wide and 0.36 metres thick. 2. Piers of Guang Liang gates are 0.75 metres high, 0.45 metres wide and 0.30 metres thick .3. Piers of Ru Yi Gates are 0.60 metres tall, 0.29 metres wide and 0.19 metres thick. 4. Gates piers of common people's dwellings are 0.59 metres tall, 0.25 metres wide and 0.19 metres thick. Gates piers of princes' mansions and common people's houses have great difference not only in carving patterns but also in carving workmanship. The bulk of bass relief carvings are flowers such as plum blossoms, chrysanthemums, orchids and green bamboos which are all China's well-known traditional flowers. Also there are carved pomegranate, bat, jade flute, grape and RuYi (a S-shaped article symbolising good luck). A design of musicians and dancers of ancient China carved in bass relief was found on the front and two sides of gate piers of a Prince's mansion in the western part of Beijing. Graceful features and vivid dance movements provide people with aethetic beauty. In some age-old alleys one can see the pattern of a bat holding an ancient coin in its mouth im-plying that good fortune is right before you for in Chinese pronounciation of "钱" Qian or coin and "前" Qian or before is homonymous.

The pattern also symbolises rolling in wealth, the good wish of house owner. If you look carefully one can still find these patterns in today's alleys. But we found out in the course of investigation that very few pairs of stone lions on gates piers were well preserved. Most of the stone lions were damaged during the cultural revolution, Some gate piers have only one intact stone lion left or their pair of stone lions are badly scarred with incomplete heads or tails. The tiny number of well-preserved carved little stone lions in Beijing alleys are treasured works of art indeed.

抱鼓石 Drum Stone

一对雕刻古老祥云图案的抱鼓石
Drum Stones with carved clouds.

雕刻石狮和麒麟图案的汉白玉抱鼓石
Drum Stones with carved stone lion and pattern of unicorn

“马上封侯”图
The design of monkey and horse meaning the future of promotion or granted title

“五福捧寿”图
The design, “Wu Fu Peng Shou” (five bats holding the Chinese character “寿” meaning longevity) .

“蝙蝠叼磬”图
The bat and Qing (a kind of music instrument used in ancient China)

“三羊开泰”图
The design of three goats meaning auspicious beginning of a new year

雕刻“招财进宝”图案的抱鼓石
Drum Stones with carved the pattern of Zhao Cai Jin Bao(it symbolizes wealth)

抱鼓石图案各异，小石狮生动可爱

There are different patterns of Drum Stones being like vivid little lion

前方组成方形花卉、动物图案的抱鼓石
The patterns are beautiful flowers and animals on the front part of Drum Stones. They are precious artworks now.

古老抱鼓石上的“麒麟图”
The pattern of “Unicorn” on the old Drum stones

抱鼓石上的“奔马图”
Drum stones with carved the pattern of “Galloping Horses”

京城仅存的一对六边形门墩
The only preserved a pair of gate piers of hexagon in Beijing

一对有着300年历史的老门墩
The old gate piers has 300 years long history

方形门墩
The square gate piers in different period

古老的吉祥人物门墩，左边童子手持莲花，右边道童捧太平花

The auspicious pattern of figure on gate piers, the pattern of a child with a lotus flower in his hand on the left gate piers, the pattern of a child holding a flower in his hand on the right gate piers

雕有单钱和双钱图案的方形门墩（又称“福在眼前”）
A square gate pier skulpted with one coin or two coins entitled, “Fu Zai Yan Qian” or happiness is at hand

“五福捧寿”图案的门墩
The design, “Wu Fu Peng Shou” or five bats holding the Chinese character “寿” (meaning longevity)carved on an age-old gate pier.

雕有龙云图案的方形门墩
The pattern of clouds and a dragon on a square gate pier

不同图案的方形门墩

The square gate piers of different rare patterns

"狮子绣球"浮雕图案，画面生动活泼，是胡同中仅存之精品

The vivid and lively carved design, "Lions playing with a ball" carved in bass relief is the only exquisite piece of art so far extant in Beijing alley

雕有小石狮的方形门墩

A square gate pier carved with little lions

方形门墩，两侧均雕刻“暗八仙”图案
The square gate piers are sculptured with designs of 8 immortals presented in a concealed way.

雕刻“佛八宝”图案

The patterns of the eight Buddhist treasures symbolized auspiciousness

“奔马图”石雕图案生动，是方形门墩中的珍品。它已存世300余年

The design, “Galloping horses”, is a precious work of art found on square gate piers. It has a history about 300 years.

143

年代久远、图案丰富多彩的方形门墩
Ancient and richful patterns of square gate piers

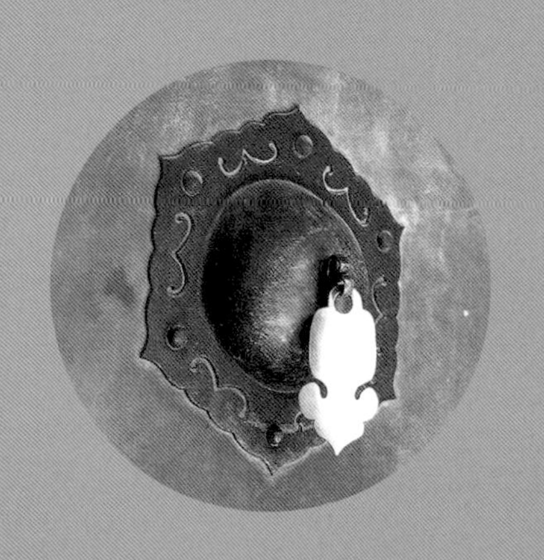

门簪、门钹、门联

Clasp, Cymbal-Shaped Knocker and Couplet

门簪 Clasp

门簪是四合院门楼大门上的木制装饰构件，同时起到加固门框的作用，广亮大门上四个门簪，如意门上多为两个门簪，迎面浮雕花卉牡丹、葫芦以示富贵之意。但雕文字的为多数，四个门簪上迎面各雕一字，如“惠我迪吉”，其意是引我来至吉祥的地方。两个门簪雕有“吉祥”、“平安”、“迪吉”、“福寿”等字样，显示宅主人的向往和愿望。

※

Clasps are decorative wooden parts on the gates of quadrangles. They also help to reinforce the gate' frames. Usually Guang Liang gates have 4 clasps and Ru Yi gates 2 clasps. They are carved with peonies or gourds symbolising wealth and nobility. However most of them are inscribed with words. On the top of each of the four clasps is a word. Four words form a phrase like “Hui Wo Di Ji” meaning leading me to the auspicious place. Words carved on two clasps imply “Ji Xiang” (auspiciousness), “Ping An” (plain sailing), “Di Ji” (propitousness) and Fu Shou (good fortune and longevity) expressing the house owners' aspirations and wishes.

门钹 Cymbal-Shaped Knocker

门钹，由铁或铜材料所制，装饰在大门的左右各一个，成对称位置，其形状类似民乐中的“钹”，称为“门钹”。在其中心装有树叶形金属片或金属圆环。自明、清两代，以至民国，用来敲门发出响亮的金属声，传至院里，当人听到便会来开门。按宅第之分，门钹造型、尺寸大小均有所区别。

Cymbal-Shaped Knockers are made of iron or copper which are symmetrically fitted on the right and left sides of the door leafs. They are so called because they look like cymbals of Chinese traditional instruments. Ever since the Ming and Qing dynasties down to the Republic of China from 1912 to 1949 they had been used to give out loud metallic sounds. On hearing the sounds people in the quandrangle would come out to open the gate. The knockers differ in design and size in keeping with the type of gate.

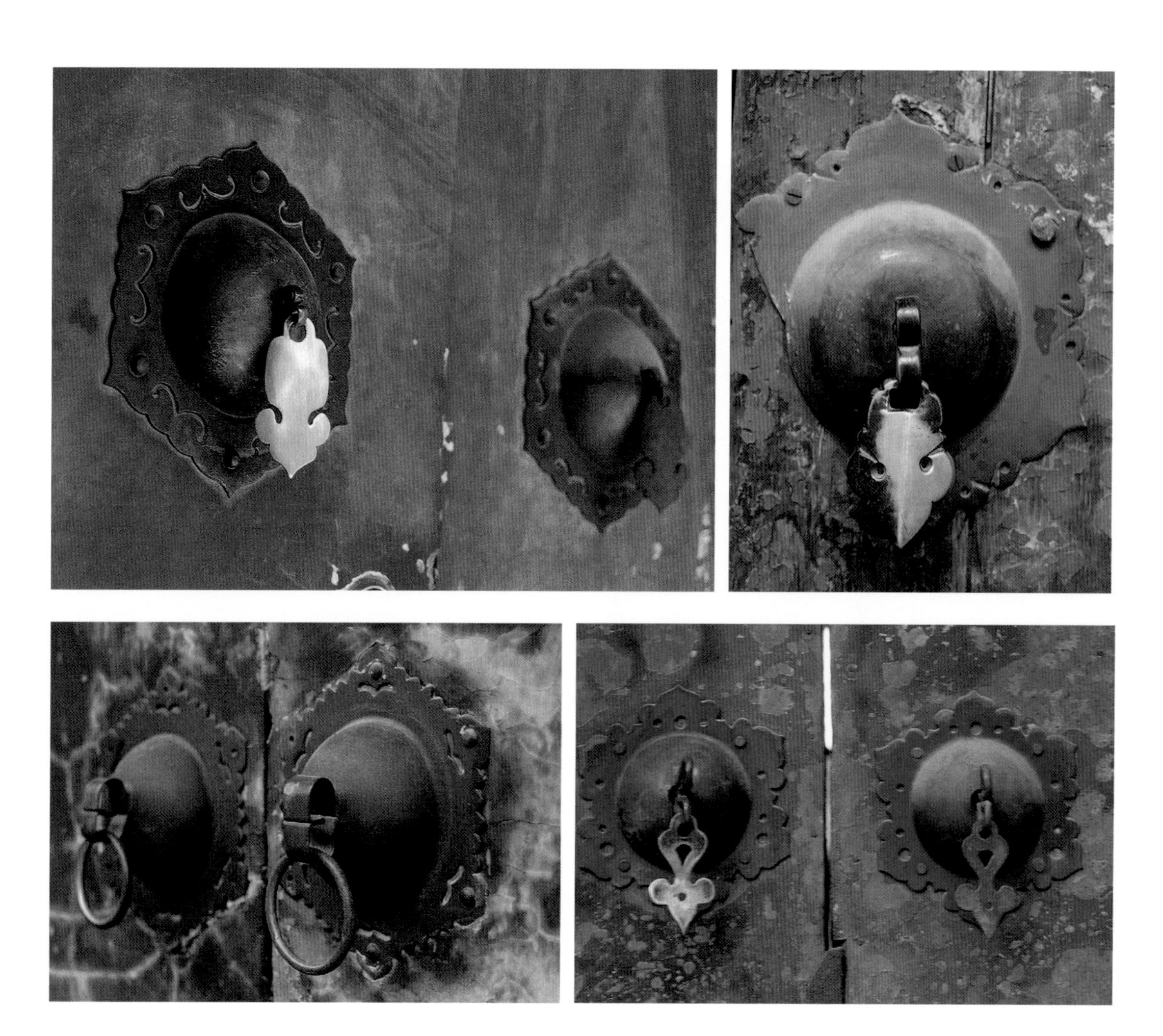

不同造型的铜质门钹
Bronze Cymbal-Shaped Knockers of various shapes

不同造型的铁质门钹
Iron Cymbal-Shaped Knockers of various shapes

门联 Couplet

门联。北京四合院门楼前不仅有雕刻精美的门头砖雕和门墩，还有大门上的门联，左边为上联，右边为下联。门联内容表达了宅主人的思想情操和对美好生活的志趣。精心雕刻在大门上的对联分别采用了隶、篆、行、楷多种书法形式。如果细心阅读胡同里的门联，则既品味了中国传统文化的内涵，又得到了书法艺术上的享受。多数门联为四字、五字、七字对句，但也有少数门联采用三字、八字对句。其内容十分丰富，有的还用黑漆描饰。三字对联如："仁由义，德载福。"四字对联如："总集福荫，备致嘉祥。"五字对联有："忠厚传家久，诗书继世长。"七字对联有："传家有道为存厚，处世无奇坦率真。"

值得一提的是下门槛。它是门框四边的组成部分，起着坚实门框、容纳门扇的功能，还起到防风、防水、防盗和阻止财气外流的作用。

In front of the gate tower of Beijing quadrangle, there are not only exquisite the brick carving on the gate head and gate pier, but also couplet on door panel. (the first line of a couplet is carved on the left door leaf, while the second line of a couplet on the right door leaf.) The content of the couplet represents the idea and sentiment of master of the residence as well as his aspiration for inviting life. Manifold calligraphy form. such as official script, seal script, cursive script and regular script was applied on the couplet, which was carved on the gate with meticulous care. If you read the couplet on door panel in the alley with circumspection, you will obtain enjoyment on the calligraphy while tasting the meaning of the Chinese traditional culture. Most of the couplet on door panel is in the form of four -word, five-word or seven-word distich, while some few in the form of three-word or eight-word distich. The content of the couplet is very abound, some of which is painted and decorated with black paint. The three-word couplet, taking for example, the first line of which is "Ren You Yi" and the second line is "De Ji Fu" .(means that Benevolence spring from righteousness, virtue brings happiness.) The four-word couplet for instance, "Zong Ji Fu Yin" written in the the first line while "Bei Zhi Jia Xiang" on the second line, meaning that inviting goodness and auspiciousness, collecting happiness and blessing. Taking a five-word couplet for example, it reads "Zhong Hou Chuan Jia Jiu" , "Shi Shu Ji Shi Chang" , meaning that loyalty and honesty carry forward family traditions long, education continues family traditions for generations to come. One seven-word couplet, the the first line of which reads "Chuan Jia You Dao Wei Cun Hou" and the second line of which reads "Chu Shi Wu Qi Tan Shuan Zhen" , means that, the way of carrying forward family traditions is none other than to be faithful and honesty, and the philosophy of life has no secrets but to be truthful and sincere.

What's worth mentioning is the lower threshold. It's a component part of the door frame. It serves to reinforce the door frame and tighten the closed door leafs as well as to prevent wind and water, robbery and run off of wealth.

门联
The couplet

25
卫生模范院
古國文明盛
新民進化多

物華民主日
人傑共和時

圆形的顶门石墩，放置在大门里，关门后放好，以保安全。

Round stone piers are placed inside the gate and are used to reinforce the closed gate to ensure safety.

方形的顶门石墩

Square stone piers are shown in the inset

不同造型的顶门石墩

Different shapes of stone piers

倒挂楣子 Dao Gua Mei Zi

倒挂楣子是步入门楼内见到的木质装饰，均为透空。

Dao Gua Mei Zi are open-piled wooden decorations under the back eaves of gate towers which have different styles.

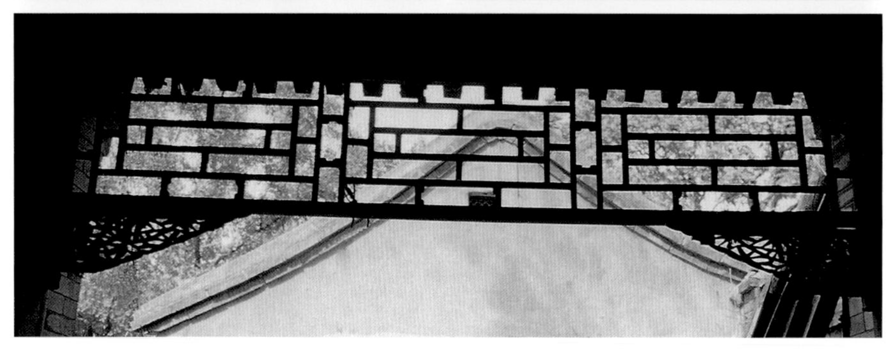

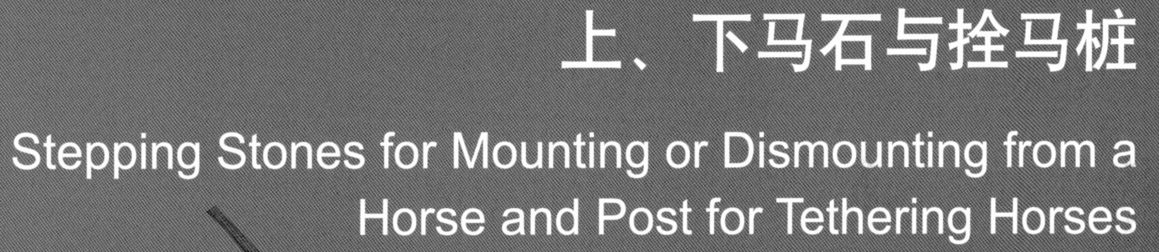

上、下马石与拴马桩

Stepping Stones for Mounting or Dismounting from a Horse and Post for Tethering Horses

在今天的国子监街（原称成贤街）路北有两座石碑，上面雕刻汉、满文“官员人等至此下马”。清朝皇家规定，满洲贵族（王、贝勒以下）年未满60岁的都骑马往返，汉族官员准许乘轿。这样京城内满族官宦人家及富户子弟出门办事都要骑马，至今一些老胡同里大四合院门前还保留着上、下马石和拴马桩。门前两侧的一对大青石，一般正面宽0.90m×0.70m或0.70m×0.60m不等，高约0.70m或0.60m，其石底部加出相连的与面宽相等的台阶，即是“上、下马石”，或称“上马石”。门第有别，规格上也有不同。王府前的上马石规格高大，正面雕有花卉，四周部分雕刻优美花纹。一般官宦富户门前的上马石，尺寸较小，仅雕光滑平面，不加图案。在门楼往西临街的“倒座”房屋墙上，均装设着四个至六个拴马桩，其式样是外周石雕方洞，里边的木柱上装有大铁环，即为拴系马缰绳所用的。今天经过走访考查，典型的大四合院，尤其是门前设有上马石及拴马桩的已极难见到了。目前，仅存的几处（指门前设有上马石及拴马桩的）四合院，也保留得不够完整，有上马石的宅院，已失去了拴马桩，有拴马桩的宅院门前，却失去了上马石。书中选了抢拍下的上马石与拴马桩实物照片，是很有研究价值的石雕珍品。

On the northern side of today Guozijian jie (street)(originally called Chengxian jie) stand two tablets on which are inscribed the words, fficials and others dismount here in the Han and Manchu languages. This re-minds us of the Qing dynasty regulations governing the imperial family. All Manchu aristocrats (below the rank of prince) under 60 in age should take horse-riding as their means of transport. Han officials were allowed to be carried in sedan chairs. Wherever they went out on errands people of Manchu official or wealthy families in Beijing had to ride on horseback. That why even today some big quadrangles in the alleys still retain stepping stones for mounting and dismounting from a horse and posts for tethering horses. A pair of large blue stone slabs are placed in front of the gate with their front sides measuring 0.90 metres × 0.70 metres or 0.70 metres×0.60 metres in area and a height about 0.70 metres or 0.60 metres. At their bases are added steps with the same width. They are stepping stones for mounting or dismounting from a horse. They differ in standard according to the rank of their master. For instance, the stepping stones for mounting or dismounting from a horse in front of princes residences are big and tall with floral carvings on the top and beautiful, exquisitely-carved designs on four sides, while those of official or wealthy families are smaller in size. Though the stone slabs are of top grade and their surfaces are polished smoothly, they have no carvings of any design. On the wall of the Dao Zuo house facing the street, there are four to six posts for tethering horses. They are wooden posts equipped with big iron hoops which set inside square holes surrounded by carved stones. Through on-the-spot investigation we could rarely find typical big quadrangles with stepping stones for getting on or off from a horse and posts for tethering horses. What the few existing ones have been preserved are incomplete, some have retained stepping stones for getting on or off a horse but lost the posts for tethering horses or vice versa. The real objects whose pictures we took in time and whose pictures were carried in this book are precious stone carvings of great research value.

北京东城国子监胡同里，刻有“官员人等至此下马”汉、满文字之石碑

In the Guozijian Street in the eastern district of Beijing stands a stone tablet inscribed with the words, “Officials and others dismount here” in Han and Manchu languages.

位于故宫东门外的下马碑

The stone tablet at the easten entrance of the Forbidden City

清代时，宅主人出门办事，由仆人牵马，主人登此石上马

During the Qing dynasty when a house owner went out on an errand he mounted the horse from these stepping stones while a servant held the reins

至今保存着有上下马石及拴马桩的一座四合院（在东城区飞龙桥胡同内）
A quadrangle at the Feilongqiao Hutong (Alley) eastern Beijing has preserved stepping stones for mounting or dismounting from a horse and posts for tethering horses.

保存完整一对上下马石的四合院门楼

A pair of stepping stones for mounting or dismounting from a horse which have been kept intact

北京尚存年代悠久的上马石近照。正面浮雕麒麟，侧面浮雕奔马图案

Time-honored stepping stones for mounting or dismounting from a horse now extant in Beijing. Bass relief carved unicorn on the front side. Bass-relief carved galloping horses on the side face.

石雕古钱图案造型的拴马桩的近照

A close-up of a post for tethering horses with a stone carving design of ancient coins. This is another form of posts for tethering horses.

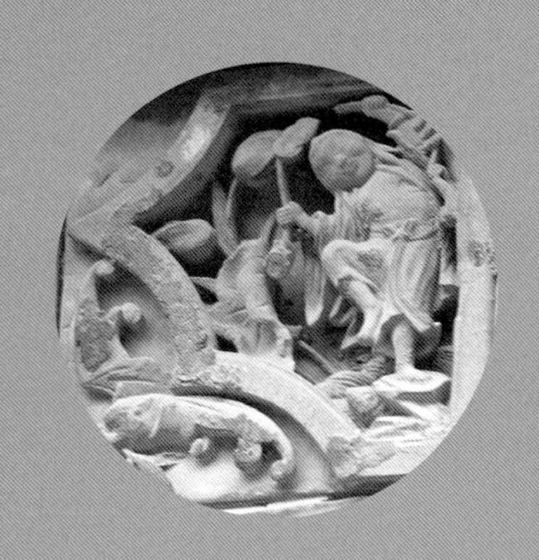

影 壁

Screen Wall

影壁，又称为照壁。其位置在四合院门楼的对面，用于遮挡对面杂乱的建筑物，使人们由院内走出大门时感到宽阔、整洁，以示内外有别。古代风水学中，认为影壁是针对冲煞而设置的。《水龙经》云：“直来直去损人丁”，古建筑设计中忌讳直来直去，故门楼前设置影壁，或院内设影壁，使气流绕着影壁而行，气则不散，符合“曲则有情”的原理。北京胡同中四合院门前的影壁由砖、石砌成，其造型规格尺寸各有不同，与宅主人的官阶、身份有关。影壁大致可分如下几类：

1. 呈“一”字形的，俗称“一字影壁”。其建筑分三段式：上有筒瓦、中壁做出仿木结构的梁框架、框心及四角加上砖雕，下有须弥座。

2. 整体造型呈“⌐”形的影壁，名为“雁翅影壁”。砌工精细，磨砖对缝，在影壁墙上边角雕有简单的花草浮雕，既庄重且美观。今天在胡同中保存完好的已经不多了。

3. 在一些古老胡同里的府第之家，宅门东西两侧，砌影壁墙与门楼檐口成130°左右的夹角，称为“八字影壁”（俗称八字粉墙）。在两侧墙面雕有对称图案的巨幅砖雕。宅门与影壁墙之间留有空地，并放置上下马石一对。

北京有些老街巷里四合院门前影壁独特而出名。东城的麒麟碑胡

Screen walls called Ying Bi or Zhao Bi or in Chinese are built facing the gates of quadrangles and used to shut out the disorderly buildings outside and to give people a feeling of spaciousness and tideness when coming out of the gates. According to the theory of geomancy screen wall is constructed to dispel evil spirits. The book Shui Long Jing of ancient China says, traightforwardness will reduce the family number" . So straightforwardness is a taboo in ancient Chinese architectural designing. That's why screen wall is built in front of the gate or in the courtyard so as to make air flow circle the screen wall in conformity with the principle, "winding arouses interests" . Screen walls facing the gates of quadrangles in Beijing alleys are generally built of brick and stone. They vary in design, standard and size according to the offical rank and social status of their owner. They can be roughly divided into the following types:

1. Those shaped like "一" are called Yi Zi screen walls. They have 3 parts: the upper part is covered with semi-circle tiles, the middle part is an imitated wooden structure with brick carvings on four corners and the lower part is the base.

2. Those shaped like "⌐" are called Yan Chi or wild goose-wing shape screen walls. They are meticulously constructed with finelyjoined polished bricks and simple bass relief carvings of flowers and grass on corners, looking sedate and elegant. Not many of them can be found in today's alleys.

3. Ba Zi screen walls, so named because they are shaped like the Chinese character "八" . They are erected on the east and west sides of the gates forming an included angle of 130° or so against the gates's eaves. Sometimes each of the two walls of these screen walls bears a similar giant brick carving. Between the screen walls and gate there are open spaces where stepping stones for mounting and dismounting from a horse are installed.

Some old Beijing alleys are famous for their unique screen walls. For instance, the Qilinbei or Unicorn Tablet Alley in the east part of Beijing. It formerly had a white marble screen wall over 2.7

同，原有座2.7m余宽的汉白玉影壁，浮雕麒麟，人面鳞甲，生有十根角，形态逼真。此物原系明嘉靖年间武将仇鸾府前影壁所嵌，后曾埋于地下，清末时出土，放置在此。胡同也因此影壁浮雕麒麟而传名。辛亥革命后原石雕移至北京的鼓楼内。

德胜门内的铁影壁胡同，原留有元代影壁一座，可算北京最古老的影壁，由一巨大火成岩雕制而成，色泽铁黑，故人们称它为“铁影壁”。现该影壁被移至北海公园快雪堂前。佟府夹道胡同，今166中学，原是清康熙皇帝内亲佟国纲、佟国维的府第。门前原有影壁一座，上镶嵌汉白玉巨石一块，高2.2m、宽0.90m。这块玉石上，有青、褐色天然花纹，雨后经水冲洗，隐约可见在山峦云雾之上，有一观音菩萨坐像，其发髻、披巾、眉眼均由天然花纹形成。坐像前边有一个香炉，香烟缭绕，很是逼真，可称为世间的奇石。这座影壁因有这观音像画石而传名。此石现保存在该校院内。

阜内大街有一座巨大影壁，高约5.5m、长约34m、厚1.3m。该址原为历代帝王庙，为宗庙建筑，虽不可与四合院并列，但此影壁却是北京现存街巷中最长的影壁了。

metre long carved in bass relief an unicorn with a human face, a body covered with scales and 10 horns. The carving was originally inlaid on a screen wall in front of the residence of Qiu Luan, a general of the Jiajing reign of the Ming dynasty. Later it was buried underground. After being unearthed at the end of the Qing dynasty it was placed in the alley, hence the Qilinbei or Unicorn Tablet Alley. The original stone carving was moved to Gu Lou or the Drum's Tower after the Revolution of 1911. The Tie or Ironcolored Screen Wall Alley inside the Desheng Gate is so named because it formerly had an iron-colored screen wall bearing a giant carving of igneous rock. Buit in the Yuan dynasty it is Beijing' s oldest screen wall. It has now been moved to the front of the Kuaixue Hall of Beihai Park. The No. 166 High School in Jiadao Alley was formerly the mansion of Tong Guogang and Tong Yuanwei, relatives on the queen's side of Emperor Kangxi of the Qing dynasty. Facing the gate was originally a screen wall set with a big piece of white marble 2. 20 metres tall and 0. 90 metres wide. After washed by rains the bluish and brown natural veins of the marble faintly present a statue of Buddha Guanyin sitting above the mountains and clouds. Buddha Guanyin' hair style, shawl, brows and eyes are all shaped by natural veins. It' s really a rare piece of marble in the world. In front of the sitting Buddha Guanyin there used to be an incense burner with curling smokes, making the Buddha Guanyin's image more vivid. Because of this, the screen wall' s name is spread far and wide. This marble screen wall is now kept. in Funei Dajie street stands a huge screen wall 5. 5 metres tall, 34 metre long and 1. 3 metres thick. It is Beijing' s largest screen wall so far extant. The site was originally the place where ancestral temples of various dynasties were built. Quadrangles cannot stand side by side with it. The photos carried in this book to which you may refer would arouse your interests to study and appreciate the screen walls in Beijing's alleys.

广亮大门前的一字影壁

The screen wall in front of a Guangliang Gate. It is popularly known as Yi Zi screen wall or screen wall in a row.

老胡同里，官宦之家四合院前的雁翅影壁

The wild-goose-wing screen wall in front of the gates of official families. Shaped like "⌒" they are meticulously constructed with finely-joined polished bricks and simple bass relief carvings of flowers on corners.

影壁及局部精美砖雕
The screen wall and the brick carving on it

走进大门里见到的精美照壁
Beautiful screen walls inside the gate of quadrangles

鴻禧

表现民间传说“刘海戏金蟾”故事的图案

The design portrays a scene from the folk legend “Liu Hai plays with a toad”.

北京市内最长的影壁（位于阜成门内大街路南）
The longest screen wall in Beijing (the south of Funei Dajie street)

位于平安大街上的另一个长影壁
Another long screen wall at the Pingan Dajie street

南锣鼓巷

Nanluogu Lane

东城区的南锣鼓巷是一条南北向的长街巷，南端是地安门东大街，北端是鼓楼东大街。在这条长巷中，东西对称的胡同有16条，连同主街形成了蜈蚣状，老百姓称之为“蜈蚣巷”。从明、清两代保留下的地图资料看，这些胡同形成于元代，经过数百年的风云变幻，今天依然保存原样；胡同的名称也大部分为老名字，其建筑风貌更保持着老街道的特点。胡同内不同类型的四合院门楼整齐，有的老宅院门前还保存着上下马石和雕刻精美的抱鼓石，这是胡同文化的实物写照，另外还有众多现存的名人故居、清代的宅第和园林，已成为文物保护单位。

清代，该地区满族镶黄旗住户居多，有官宦之家，也有众多的小户民居。南锣鼓巷主街上商店众多，饽饽铺、茶叶铺、粮店、酒店、当铺、饭馆、猪肉扛、菜店……可谓店铺俱全。胡同内均为民居，从未开设商户，百姓们在此地区生活得安宁和谐。改革开放以来，自20世纪90年代中期，南锣鼓巷出现了各类商店，如有京城特色的奶酪店、书店、酒吧等。胡同旅游的三轮车来到了老街巷，众多的中外旅游者对这里的四合院和民俗深感兴趣，在街巷中留下他们的足迹。

Nanluogu Lane is a south-north lane which links the east Avenue of *Di'an* Gate at the south end and the east Avenue of Drum Tower at the north end. Sixteen *Hutongs* locates symmetrically at the east and west which constitutes a map of scolopendra and that is why the lane is called the scolopendra lane by the locals. According to the records of maps of *Ming* and *Qing* Dynasties, the *Hutongs* came into being in *Yuan* Dynasty but remained its origin through the long history. Even the name of the *Hutongs* keeps it old call, no matter the feature of the old streets. All the diversified quadrangles are in order, in front of some of which there are stepping stones and fine works of drum piers. As a historical portraiture of *Hutong* culture in Beijing, you can also find many cultural relics such as the former residence of celebrities, mansions and gardens in *Qing* Dynasty. In *Qing* Dynasty, most of the residents of this area are *Xianghuangqi* of Manchu, including both rich merchants and the civilians. Almost all kinds of shops can be found in the *Nanluogu* Lane, such as bun shop, tea shop, foodstuff shop, groggery, pawnshop, restaurant, pork store and vegetable shop. However, inside the dozen of the *Hutongs*, there has never been any store and only the populace lived here in peace and harmony. Since the Reform and Open movement, especially after the middle of 1990s, *Nanluogu* Lane is enriched by various stores such as famous Beijing cheese store, bookstore and pub. The pedal car shows the foreign visitors around, down into the quadrangles and the national culture and customs are extremely attractive to the tourists.

奎俊府 Mansion of *Kui Jun*

位于黑芝麻胡同13号，是清代四川总督奎俊的府第，共有五进院落，是典型的多进大四合院。门楼前有上马石一对，大门洞内有抱鼓石一对，其雕刻古朴大方。院里的垂花门、抄手廊都很具特色。现花园部分为黑芝麻胡同小学，住宅部分为东城区保护文物。

※

This is the residence of *Sichuan* viceroy *Kuijun, Qing* Dynasty. There are two mounting stones in front of the gate and a pair of Drum Piers inside the gate, the carving of which is simple but elegant. Quinary quadrangle as it is, this mansion is typically a huge multi-tier courtyard. The festoon door and corridors are both designed finely. Now, it is a bachelor hall.

总督府

the Governor's Mansion

文煜故居 The Former Residence of *Wenyu*

位于帽儿胡同11号，是清代大学士文煜的宅第。这是一座五进院落，布局严谨，屋宇高大，庭院宽敞，原为文煜所建。北洋政府时期冯国璋居此。新中国成立后一度用作朝鲜大使馆，现为东城区文物保护单位。

※

This is the residence of Great Scholar *Wenyu*, *Qing* Dynasty. This is a quinary quadrangle which is tall in room-design and wide in courtyard. It was built by *Wenyu* first and then came to be the residence of Feng Guozhang in *Beiyang* times. After liberation, it was the embassy of North Korea. While now, it is on the list of relic preservation.

文煜故居

The former residence of *Wenyu*

婉容故居 The Former Residence of *Wanrong*

位于帽儿胡同35号，是清朝末代皇帝溥仪的皇后婉容的旧居，原为婉容曾祖父郭布罗长顺所建。现西院建筑完好，四进院落；东路为三进，有月亮门、假山、祠堂等；后院正房三间，室内还保存着一面巨大的镜子，是清代由国外购置的。婉容在入宫之前，每日在镜前演习礼仪。今天这面巨镜已成了文物。

This is the former residence of Empress *Wanrong* of the last Emperor *Puyi* of *Qing* Dynasty, which was built by the great-grandfather of *Wanrong*, *Guobuluochangshun*. When she was selected as the empress, her father was promoted as the minister of Imperial Household Department, a 3rd rank *Cheng'en-gong*. The two yards are well-kept today. The east route is ternary with Moon Gate, rockery and ancestral temple, while there are three main rooms at the back yard, inside which a huge mirror imported from the abroad was put. Before *Wanrong* was selected, she practiced royal manner in front of this mirror. And today, the mirror has become a cultural relic.

婉容故居
The former resideme of *Wanrong*

婉容故居正房中精美的木雕落地罩
Delicate wooden floor-lamp cover in the main room of the former residence of *Wanrong*

文昌帝君庙 The Temple of Emperor *Wenchang*

在帽儿胡同西口路北小巷内，有个老院子，原是明代成化十三年（1477年）始建的文昌帝君庙，百姓称它为“梓橦庙”。原建有山门、钟鼓二楼、正殿，正殿内供文昌帝君像。文昌帝君是主管文运之神。现在仅存大殿，为居民所住，院内保存着一座巨碑，通高4米，上有清嘉庆皇帝撰文、乾隆年间大学士刘墉所书碑记，记述该庙的历史。石碑历经200余年仍保存完好，尤为珍贵。近前观赏，碑上所刻的楷书字迹清晰可辨。

※

Down to the lane in the north road of the west entrance of *Mao'er Hutong*, there is an old courtyard which is the temple of Emperor *Wenchang* first built in the 13th *Chenghua* Year in *Ming* Dynasty (1477 A. D.). The local people call it the *Zitong* Temple. There used to be a frontispiece, a drum tower and bell tower with the statuary of Emperor *Wenchang*, the God in charge of the achievements in culture and education. However, the only kept main palace has turned out to be the residence of the locals. But there is a huge stele standing in the middle of the yard which is four meters in height. The words on the stele tell the history of the temple which is initiated by Emperor *Qianlong* and written by the Scholar Liu Yong. This stele is of great value and importance as a cultural relic since it is wll-reserved after the baptism of over two hundred years. The words carved on the stele can even be clearly seen with a close look at. The script is the autography of the Prime Minister Liu Yong, who is more and more popular recently by means of the series plays.

文昌帝君庙院内保存完好的石碑

Well-kept stele inside the yard of the temple of Emperor Wenchang

齐白石故居 The Former Residence of *Qi Baishi*

雨儿胡同13号，原为清代内务府总管大臣的宅第。新中国成立后，著名国画家齐白石曾住在这里，后在此建立齐白石纪念馆。该院为二层大四合院，其建筑很有特色。进里院带转角廊，此屋明间木隔扇上刻有对联：

本书以求其质，本诗以求其情，本礼以求其宜，本易以求其道；

勿展无益之卷，勿吐无益之话，勿涉无益之境，勿近无益之人。

横额：

乐生于智，寿本乎仁。

此等佳句可使人们读后深思。现在，该处为北京画院使用。

No. 13 of *Yu'er Hutong* is the former residence of general minister of Imperial Household Department. Famed artist Qi Baishi lived here once after 1949 and then it is turned to be the Memorial to Qi Baishi. This is a binary quadrangle which is different in the crossing corridor in each yard. There is a couplet on the wood door: *Reading books for its connotation, enjoying poems for its passion, behaving polite for fine relation, being simple for its Tao; rejecting the worthless books, dumbing for meaningless words, removing from the profitless environment, keeping aloof from the virtueless people.* The banner in the middle is *Happiness derived from wisdom, longevity rooted in benevolence.* The sentences do make people deliberate. Today, the house is used as a gallery.

齐白石故居

The former residence of *Qi Baishi*

茅盾故居 The Former Residence of *Mao Dun*

位于后圆恩寺胡同路北，是一座三进的清末老四合院。茅盾，原名沈德鸿，字雁冰，浙江桐乡人。中国现代进步文化的先驱，伟大的革命文学家，卓越的无产阶级文化战士。他从1916年开始从事文学活动，对我国新文化运动产生了巨大影响。新中国成立后，任文化部部长、全国文联副主席。茅盾创作了《子夜》、《蚀》、《春蚕》、《林家铺子》、《虹》等文学作品。自1974年12月茅盾与其子居住在这里，直至1981年3月病逝。在这个院子里，有先生生前的起居室和工作室，陈设全是旧物，保留原貌。在前院立塑像，辟有展室，举办“伟大的革命文学家——茅盾”展览。

The house locates in the north road of the Temple of *Houyuan'en*, a trinary quadrangle of the late *Qing* Dynasty. Mr. Mao Dun, the original name of whom is Shen Dehong was born in Tongxiang, Zhejiang province. He is a pioneer of Chinese modern culture, a great revolutionist and litterateur and an outstanding proletarian soldier. Mr. Mao devoted himself into cultural activities in 1916 and had great influence on national new cultural movement. After the founding of PRC, he worked as the minister of national Ministry of Culture and the vice president of China Federation of Literature and Art Circles. He produced the famous works such as *Midnight, Erosion,*

Spring Silkworm, the *Store of Lin's* and *the Rainbow*. Mr. Mao and his son have lived here since December, 1974 and not until March, 1981 that Mr. Mao passed away here. The living room and workroom are still kept in order with the old staff in exhibition. The statuary of Mr. Mao Dun was erected in the front courtyard and the special exhibition room for him is open to the pulic.

茅盾故居

Former residence of Mr. *Mao Dun*

茅盾故居院内一景

Spot in the courtyard of the former residence of *Mao Dun*

起居室一角

Corner of the living-room

可园 Ke Garden

位于帽儿胡同，是清末大学士文煜的旧宅花园。原主人的住宅与花园相通，现已隔开。花园部分很美，进门经过弯曲小径，直通假山，山洞上刻“幽径”二字。走过山洞，眼前豁然开朗，往北可见正房五间，往东过小桥又到亭台，此处较高，可放眼全园。太湖石旁有石碑一座，上刻“可园”二字。游廊依地势而建，可通向后园。园内四季树木，花草丛丛，甚是优美典雅。现为北京市级文物保护单位。

※

This is the garden of the old mansion of the Great Scholar Wenyu, *Qing* Dynasty. It was originally linked with the mansion, but was separated now. Along the winding paths, you come to a rockery with words *Youjing* carved on the cavern. Going through the cavern, you will have a wide vision at once. There are five main rooms to the north and a pavilion across a bridge to the east. The pavilion stands at a higher position where you can have a outlook of the whole garden. By the *Tai* Lake Stone, there is a stele carved *Ke* Garden. Corridors built delicately according to the change of the hypsography, together with the green trees and blooming flowers, turns out a nice picture. It is also one of the preservation of cultural relics.

可园一景

Spot of the *Ke* Garden

板厂胡同老宅

Quadrangle in No. 27 *Banchang Hutong*

位于板厂胡同27号，是一座保存完好的三进四合院，门内有砖雕影壁，外院倒座房六间。走进垂花门，呈现抄手游廊；正房三间，左右各有耳房两间；经月亮门可转至后院，后罩房七间。此宅院为清代建筑，很有晚清庭院特色。

※

This is a well-kept ternary quadrangle. There are a brick carving screen wall inside the gate and six inverse side-rooms in the outer yard. Stepping into the festoon door, there is the *Chaoshou* Corridor. By the two sides of the three main rooms, there are two ear-rooms. You can get to the back yard through the Moon Gate and there are seven *Houzhao* rooms there. This is a *Qing* building and even today, we can still find the feature of the courtyards of late *Qing*.

板厂胡同27号四合院

Quadrangle in NO.27 *Banchang Hutong*

荣禄故宅　The Resident of *Ronglu*

位于菊儿胡同3号旧宅院，是清光绪时直隶总督荣禄的宅第。大门坐北朝南，院内月台上正房五间及东西厢房均为清代建筑，至今保存完好。

※

The residential No.3 in *Ju'er Hongtong* is the mansion of Zhili Viceroy, *Ronglu*, in *Guangxu* time of *Qing Dynasty*. It's on the north and to the south. The five room on the platform in the yard and the west and east side-room are *Qing* buildings and are kept well.

荣禄故宅
The resident of *Ronglu*

蒋介石行辕

Quadrangle in No.7 Houyuan'ensi *Hutong*

位于后圆恩寺胡同7号，此院原是清代庆亲王奕劻次子载旉的宅第。除正规的老四合院建筑之外，还有一座西洋式的小楼房和花园。抗战胜利后，这里是蒋介石的行辕。新中国成立后，该处为中国共产党华北局所在地。今天此院为友好宾馆，迎来了大批中外旅游者。

※

This is the mansion of the second son of Prince *YiKuang, Qing* Dynasty. The whole building has an old quadrangle as well as a western-style storied building and a garden. After the success of the anti-Japanese War, it became the xanadu of Jiang Jieshi. When China was liberated by the CPC, it came to be *Huabei* Bureau of CPC. Today, it is *Youhao* Hotel which has been open to tourists around the world.

蒋介石行辕部分景观
（刘松年摄）

Part of the temporary residence of *Jiang Jieshi* (photographed by Liu Songnian)

地安门古桥 *Di'an* Gate Bridge

该桥在地安门外大街上，始建于元代，名为万宁桥。因地安门又称后门，因此称为后门桥。近年进行了全面修缮，是一座单孔石桥，桥下岸边有镇水兽一对，石雕精美壮观。

※

The bridge locates on the street of *Di'an* Gate, first built in *Yuan* Dynasty. The original name of the bridge is *Wanning* bridge. Since *Di'an* Gate is also called the Back Gate and the bridge can also be called the Back Gate bridge. The singe-hole stone bridge has been completely restored recently. Under the bridge, there are a pair of protection beasts whose carving technique is finery and brilliant.

古老的后门桥

Old Back-door Bridge

索家花园 The Garden of Family Suo

此园在秦老胡同35号。原为清代内务府总管大臣索家的花园，名为“绮园”。院面积2000余平方米，除假山、水池、亭台等建筑外，还有一仿江南园林中的船形敞轩。形似大船，造型奇异，是很有特色的园林。

※

The garden of Qi yuan lies in No.35 QinlaoHutong. In the past, it used to belong to ower of family Suo who was the minister general manager of imperial palace in Qing Dynasty. There are the Chinese Jiangnan-style(south of the yangtze River) spacious pavilion, the artificial hill, pond, pavilions in its floor space of 2000 square meters.

索家花园正门

the front door of the Garden of Family Suo

万庆当铺 Old Pawnshops

在南锣鼓巷主街路东，至今保留着始建于清末的万庆当铺，砖墙门楼下雕有“萬庆”二字，典当业盛行在清同治、光绪年间。店铺深墙后壁用以防盗、避火灾。现在此遗址保护较好。

The Wan Qing pawnshop stands in the east side of Nanluoguxiang alley, with decoration of Chinese character‘萬庆’, surrounding tall walls for fire prevention and guard against theft。It is well preserved now. The pawnshops were good business in the period of Tongzhi and Guangxu, Qing Dynasty.

万庆当铺
Wan Qing Pawn shop

御河庵 The Nunnery of Yuhe

此庵位于福祥胡同西口，明代古御河畔，是座有悠久历史的尼庙，现保存着古建和石碑，是河道旁的一处佳景。

※

The nunnery of Yuhe lies in western entrance of Fuxiang Hutong near by ancient river dating from Ming Dynasty. It is a well-preserved nunnery with a long history as a nice scene in which there are ancient sacred buildings and stone tablet inside .

御河庵

The nunnery of Yuhe

北京水准点石碑

Stone Tablet of Shuizhundian

北京的水准点是计算测量北京地区海拔高度和地形、各种建筑材料、底下构筑物的控制点。在南锣鼓巷主街路西，有个北京水准点石碑的文物，至今保留完好。这是民国6年北京市政府公布的刻石，当时北京城内共立87个，现多已毁坏，南锣鼓巷这一石碑保留完整，难能可贵。

※

Stone tablet of Shuizhundian used to be one of the positions of reference points for the height above sea level in Beijing city during 1914-1916, the period of public of China. It is said that the stone tablet of Shuizhundian here used to be the position of reference point for the highest height above sea level in Beijing city in that time. There is well preserved stone tablet of Shuizhundian at the west of Nanluogu Lane. So it is a precious one today.

南锣鼓巷主街路西的北京水准点石碑
the Stone tablet of Shuizhuidian at the west of Nanluogu Lane

胡同明珠——南

前鼓楼苑胡同 老宅院

黑芝麻胡同 奎俊府

沙井胡同 老宅院

地安古门桥

景阳胡同

文昌帝君庙碑 帽儿胡同 婉容故居 文煜故居 可园

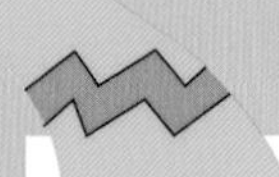

值年旗衙门 雨儿胡同 齐白石故居

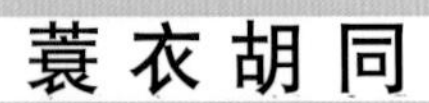

蓑衣胡同

原东部压桥遗址

福祥胡同 老宅院

御河庵

鼓巷地区览胜全图

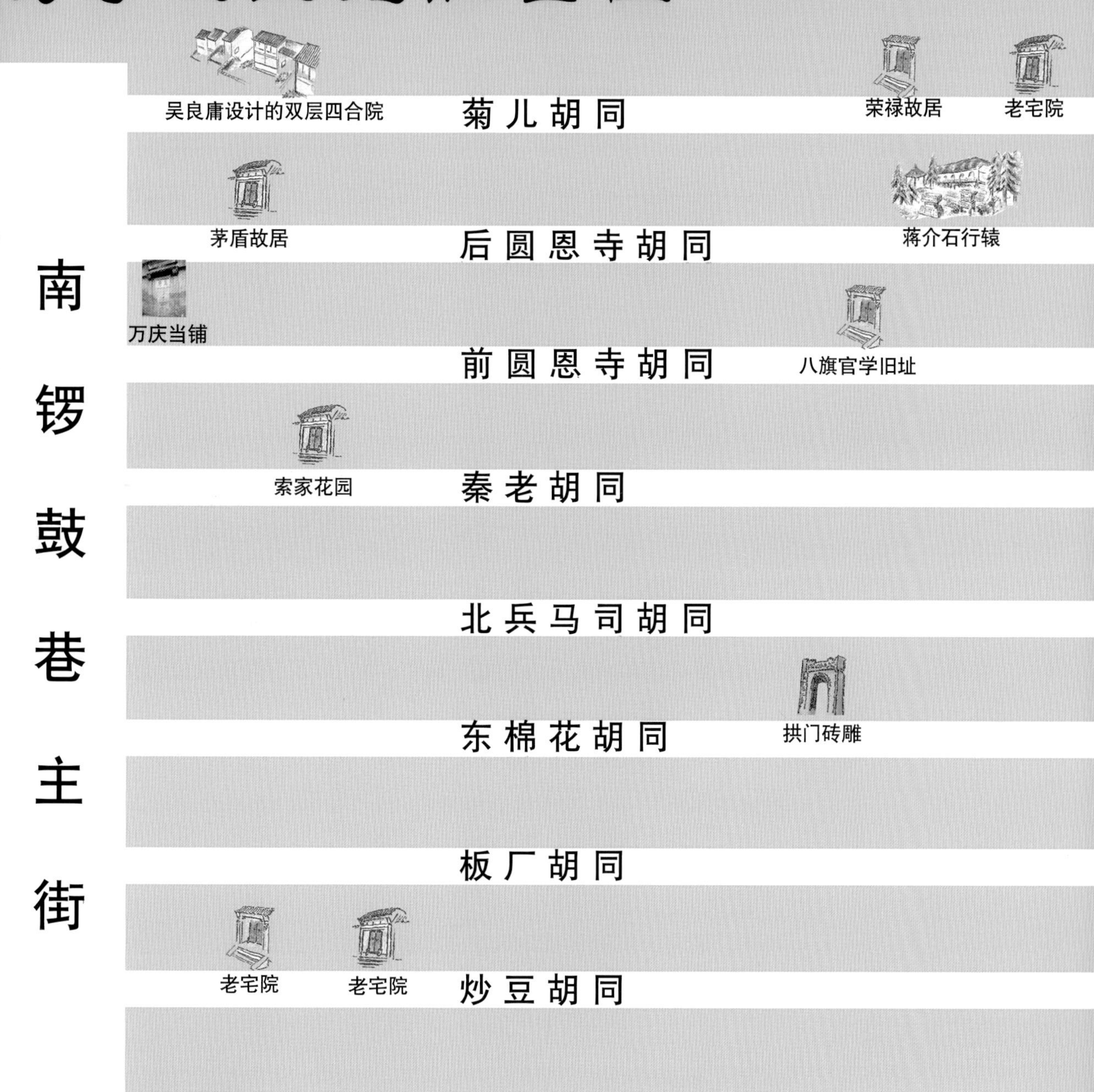

门东大街

地铁站

附录

“雕花儿匠”邓久安老人访问记

为了研究、考证北京砖雕艺术的历史和传承，我在2002年6月曾拜访过，故宫博物院古建队退休的邓久安老师傅，他是京城仅存的老技工，在古建行业中被称为“雕花儿匠”。邓师傅家住西四北二条14号，那年老人已76岁了，高挑个儿，身体很硬朗。他边向我介绍着自己的从业经历，边拿出来干活儿用的多件工具，以及保留在身边的花样子。

聊起京城的砖雕，邓老打开了话匣子：“北京有句老话：‘有名的胡同三千九，无名的小巷赛牛毛’。四合院的门楼造型多样，广亮大门门楼戗檐上雕刻的狮子、麒麟、牡丹、兰花，可是砖雕的亮点，吃功夫。还有，您看那些为数众多的如意门门楼门楹上三米长的大幅透雕，画面宽阔，内容丰富，‘多宝阁’、‘吉祥鸟兽’、‘故事传说’可是题材多样啊！都是出自历代雕花儿匠之手，这可真是门儿手艺。至

Appendix

The Visit Record to a Craftsman of Brick Carving

In June 6th 2002, I visited Deng Jiuan ,the craftsman of Brick Carving who had retired from the Palace Museum in Beijing. As the only skilled worker who was still alive in Beijing at that time, he was called as 'master of brick carving' . Mr.Deng had been living at No.14 yard in the Bei Ertiao Hutong, Xi Si area in Beijing. He was tall and looked healthy at the age of 76 years old that year. The interview was a conversation in which I put some questions to him such as his career history, and he introduced some different tools of brick carving with himself,and then told us many knowledge about the decoration of brick carving of gate building in Beijing Hutongs.

'There is a vivid saying in the past, 'the number of well-known Hutongs is 3900, and unknown Hutongs (old alleys) are countless just like cow hair.' Si He Yards are typical examples of ancient Chinese residential buildings. The styles and standards of their gates vary in keeping with the social status of their owners.' said Deng Jiuan, 'In the feudal society the residence gates demonstrated differently by the ranks and social classes of their master. They were differentiated in constructional standard and sculptural decoration. The style of Guang Liang gate buildings are more bigger and taller than common people's residence gate .Their lintels and eaves are decorated with exquisite brick carvings like Chinese unicorn, peony, orchid, etc.

今有些老门楼上的砖雕还保存得很好，这与早年间烧制出的澄浆泥砖的质量是分不开的。生产此砖最好的是山东临清，现在到哪儿去找做得如此好的砖儿啊！大面积的砖雕要用拼接的办法，分四拼、六拼、八拼来完成整个门头的图案，先定出完整的花儿样子，干起活儿来就靠手艺了。”边说着，邓师傅由工具袋里拿出十多件铁制刀具，又让我们看他保留着的花卉小样。

邓师傅接着讲：“干我们这行心要细，对绘画得懂，还要有悟性。一张草图，在砖上要变成立体的。那狮子、花草还得活灵活现，要用刻线、阴雕、镂雕、浮雕、透雕多种技巧，一刀下去就得准，一般雕完不能再打磨。我是故宫古建队退休的，我们这行俗称‘雕花儿匠’，比一般工人的待遇高，工作时间每天是6小时。我还得给你们说说这行的历史。在清乾隆年间，京城盛行砖雕，当时有位出名的‘花儿匠’，这是这行的简称，名叫李永福，他在这行里手艺最精，线刻、透雕、设计花样子都有绝活儿。他手下收了不少徒弟，京城百姓称其为‘花儿匠李’。建四合院的、修缮园林的都请他带的这伙儿工匠去干。嘉庆时期，砖雕这行更为兴盛。河北、天津的工匠也不来

Ordinarily medium-sized and small quadrangles have Ru Yi gates. Their number is the greatest in all Beijing Hutongs in Qing Dynasty. The carved brick patterns on the lintels of the gate building which are three meters long are different contents, designed as patterns of auspicious birds and other animals, folk tales. They were designed and carved by the brick carving masters. Besides the masters handcraft kills, it is very important thing that all the elaborate decorations on the brick carving consist of the materials of bricks which made of high quality mud. In the past time, these bricks with good quality which be used as brick carvings were made in Linqin area, Shandong Province. The whole carved brick decorations of the gate heads were accomplished by joining together, 2, 4, 6, 8 and more pieces. It depended on the craftsmanship of artisans to complete the patterns of figurine, flowers, etc. Once carved, usually no further polishing was needed. 'said the master Deng Jiuan. Then he showed over ten different tools for carving to me and said again, ' As a master of brick carving, it is necessary to design and carve very carefully. Besides this, you must learn about how to draw and paint the pictures. Sketches are different from the three-dimensional carving. The brick carving master used various highly difficult carving skills such as thin carving, bass relief carving ,open work carving and linear carving for creating vivid patterns of lions or flowers, etc. As usuall, they must carved exactly according to baked bricks size and sketches. I spend about six hours a day to work with higher salary than construction workers at the apartment of ancient architecture construction crew of the Imperial Palace in Beijing before retiring from there.' He continually said, ' In Qing Dynasty Qianlong period, it was popular to mount brick and stone carvings on the lintels, piers, and roof ridges of the gates of quadrangles of civilian dwellings. During that time, there was a famous the master of carving named ' Li Yongfu' who was most good at various styles of brick carvings and stone carvings, and took

作者和邓久安老人（左一）的合影

学，这就形成了京城独特的砖雕手艺。我干这行是家传，家父就是李派的传人。我读书到十六岁，后来就跟着父亲干活学艺，走上了‘花儿匠’这行。在故宫古建队就干了近五十年。”

邓师傅退休后还被大三元酒店、颐和园等处邀请，雕刻过几件大活儿。大三元酒店门前八字粉墙上保留的一对巨幅花卉砖雕，就出自邓师傅之手。邓久安老人收过两位弟子，但他们最后都没能从事这个行业。近些年砖雕这门艺术已不被人们重视，学了这行手艺，能用上的机会不多，所以愿吃苦、肯用心学并继承此行的人太少了。今天，京味砖雕艺术和“雕花儿匠”的手艺，应该列入非遗项目，才有望得到重视和传承。

many apprentices. Common people in Beijing city regarded him as ‘Hua Jiang Li’ (the master of carving Li).’ ‘Many owners of big quadrangle yard always hired Hua Jiang Li and his apprentices to design ,carve and repair the brick carvings on the gate lintels (or gate head) and Qiang Yan or eaves of the quadrangle yard. It was more popular in Jiajing period than Qianlong to carve bricks for decoration of the gate building. So it is nature to develop as the special handicraft of brick carving in Beijing. The handicraft of brick carving was handed down from the older generations of my family. My father was the apprentice of Hua Jiang Li (brick carving master Li) .I have been working for nearly fifty years at the Beijing imperial palace.’ said Deng Jiuan.

We knew from what Deng Jiuan said that he was invited by the Beijing Da Sanyuan Hotel and the Summer Palace to instruct the decoration of Brick carving. He had ever taken two apprentices to learn the handicraft of brick carving , but both of them did not do the job in the field of carving. In recent years, people have not paid more attention to the handcraft skills of brick carving. There are less and less young people who devote themselves to it. Today, it is necessary that the Beijing brick carving handicraft should be was listed in the national nonmaterial cultural heritage list for transmission of the handicraft.

后记

北京胡同的四合院门楼建筑是非常独特的。门楼造型、砖雕、木雕、石雕装饰诸方面都有着研究和观赏价值。经历元、明、清至今，北京仍保留着大量的胡同以及一座座优美的、建筑风格各异的四合院门楼。

北京这座古城，在现代化高速发展的时刻，老胡同地区在进行危改和拆迁。本书的大量照片及插图对研究胡同建筑文化有着一定的参考价值，其中的部分照片可谓精品。

Postscript

The construction of quadrangle gate building in the Beijing Hutong (alley) boasts their special originality. There are rich value of study and appreciation embedded in their shape, brick carving and stone carving decoration. After undergoing the dynasty of Yuan, Ming, Qing, a large number of has been reserved to the present. The delicate gate building, having experienced many years of swift changes of the world, still preserve a multitude of fine art article. Beijing, the historic city experiencing modernization high-speed development, is witnessing its aged alleys and region under reconstruction of old houses and rehousing. The volume of photos and a few illustrations released in the book has reference value for study of the construction and cultural of Hutong, some of which can be called collector's item.

图书在版编目（CIP）数据

胡同门楼建筑艺术／李明德著. —（增订版）. —北京：中国建筑工业出版社，2015.2
ISBN 978-7-112-17668-7

Ⅰ. ①胡… Ⅱ. ①李… Ⅲ. ①民居－建筑艺术－北京市－图集 Ⅳ. ①TU241.5-64

中国版本图书馆CIP数据核字（2015）第012875号

责任编辑：王晓迪　时咏梅
书籍设计：张悟静
责任校对：陈晶晶　刘梦然

胡同门楼建筑艺术（增订版）
李明德　著
李海川　摄影

*
中国建筑工业出版社出版、发行（北京西郊百万庄）
各地新华书店、建筑书店经销
北京锋尚制版有限公司制版
北京方嘉彩色印刷有限责任公司印刷
*
开本：787×1092毫米　1/20　印张：10　字数：300千字
2015年10月第二版　2015年10月第三次印刷
定价：88.00元
ISBN 978－7－112－17668－7
（26853）